Structure and Dynamics of Cosmos Theory and the Unified SuperStandard Theory

Dirac Equation Expanded to Generation, Layer, and Connection Interactions
Normal vs. Dark Matter Explained and Quantified in Our Universe
Generation, Layer, and Connection Groups Explained in Detail
Interaction-based Periodic Table of Fundamental Fermions
Unification of Cosmos Theory Spaces and Symmetries
Natural Quadplex Formalism for Cosmos Theory
UST Mechanism Explaining Dark Matter/Energy
Universe Creation from One Particle
UST Bradyon-Tachyon Formulation

Stephen Blaha Ph. D.
Blaha Research

Pingree-Hill Publishing
MMXXIV

ISBN: 979-8-9894084-1-2

To Margaret

Some Other Books by Stephen Blaha

SuperCivilizations: Civilizations as Superorganisms (McMann-Fisher Publishing, Auburn, NH, 2010)

All the Universe! Faster Than Light Tachyon Quark Starships & Particle Accelerators with the LHC as a Prototype Starship Drive Scientific Edition (Pingree-Hill Publishing, Auburn, NH, 2011).

Unification of God Theory and Unified SuperStandard Model THIRD EDITION (Pingree Hill Publishing, Auburn, NH, 2018).

The Exact QED Calculation of the Fine Structure Constant Implies ALL 4D Universes have the Same Physics/Life Prospects (Pingree Hill Publishing, Auburn, NH, 2019).

Passing Through Nature to Eternity ProtoCosmos, HyperCosmos, Unified SuperStandard Theory (Pingree Hill Publishing, Auburn, NH, 2022).

HyperCosmos Fractionation and Fundamental Reference Frame Based Unification: Particle Inner Space Basis of Parton and Dual Resonance Models (Pingree Hill Publishing, Auburn, NH, 2022).

The Cosmic Panorama: ProtoCosmos, HyperCosmos, Unified SuperStandard Theory (UST) Derivation (Pingree Hill Publishing, Auburn, NH, 2022).

God and and Cosmos Theory (Pingree Hill Publishing, Auburn, NH, 2023).

Newton's Apple is Now The Fermion (Pingree Hill Publishing, Auburn, NH, 2023).

Cosmos Theory: The Sub-Particle Gambol Model (Pingree Hill Publishing, Auburn, NH, 2023).

Cosmos-Universe-Particle-Gambol Theory (Pingree Hill Publishing, Auburn, NH, 2024).

Fractal Cosmos Curve: Tensor-based Cosmos Theory (Pingree Hill Publishing, Auburn, NH, 2024).

The Eternal Form of Cosmos Theory Third Edition (Pingree Hill Publishing, Auburn, NH, 2024).

Fundamental Constants of Cosmos Theory and The Standard Model (Pingree Hill Publishing, Auburn, NH, 2024).

Geometric Cosmos Geometric Universe (Pingree Hill Publishing, Auburn, NH, 2024).

Particles and Universes of Cosmos Theory (Pingree Hill Publishing, Auburn, NH, 2024).

Unification of the Subluminal and the Superluminal in Cosmos Theory (Pingree Hill Publishing, Auburn, NH, 2024).

The Dawn of Dynamic Cosmos Dimension Arrays (Pingree Hill Publishing, Auburn, NH, 2024).

Available on Amazon.com, bn.com Amazon.co.uk and other international web sites as well as at better bookstores.

CONTENTS

INTRODUCTION 1

1. SPACE AND SPIN 3

2. STRUCTURE AND SUBSTANCE 5

    2.1 STRUCTURE – DIMENSION ARRAYS 5
    2.2 SUBSTANCE - UNIVERSES 5
    2.3 SEPARATION OF STRUCTURE FROM SUBSTANCE 5
    2.4 SPACE = DIMENSION ARRAY 6
    2.5 HOW DOES THE OVERALL STRUCTURE OF COSMOS THEORY ARISE? 6

3. OVERALL STRUCTURE OF COSMOS THEORY 9

    3.1 COSMOS THEORY SPACES 9
    3.2 THE CAUSE OF DARK MATTER/ENERGY IN COSMOS THEORY AND UST ... 10
        3.2.1 Visibility of Dark Matter and Energy 10
        3.2.2 Dark ElectroWeak Gauge Theory 10
        3.2.3 Interpenetrability of Normal and Dark Matter 11
        3.2.4 Connection Group and Interactions between Normal and Dark Matter 11
    3.3 THE RELATIVE DISTRIBUTION OF THE NORMAL AND THE DARK IN OUR UNIVERSE ... 11
        3.3.1 Equipartition Principle for Particle Degrees of Freedom 11
        3.3.2 Proportion of Dark Mass-Energy in the Universe 12
        3.3.3 Proportion of Dark Mass-Energy in the Universe 12

4. BASIS OF COSMOS THEORY 19

    4.1 TENSOR BASIS 19
    4.2 DIMENSION ARRAYS 20
    4.3 CONSISTENCY CONDITION FOR TEN HYPERCOSMOS SPACES 21
    4.4 EXTENSION OF COSMOS SPACES TO FRACTIONAL SPACE-TIME DIMENSIONS ... 23
    4.5 THE FRACTAL COSMOS CURVE 24
    4.6 EUCLIDEAN CONSTRUCTION OF CREATION IN COSMOS THEORY 25
    4.7 ANALOGOUS Γ-MATRIX FEATURES 25

5. COUPLING CONSTANTS AND MASSES 27

    5.1 COUPLING CONSTANT VALUES 27
    5.2 ORIGIN OF THE FORM OF COUPLING CONSTANTS 28
    5.3 INCLUSION OF GRAVITATIONAL COUPLING CONSTANT G 28
    5.4 FERMION MASS SEQUENCES 29
    5.5 TWO SEQUENCES OF FERMION MASSES 30
    5.6 PARTICLE INTERNAL STRUCTURE BASED ON SEQUENCES 32
    5.7 DERIVATION OF THE FORM OF INTERNAL FERMION MASSES 34

6. DYNAMICAL ANALYSIS OF DIMENSION ARRAYS 37

    6.1 THE COSMOS DIMENSION ARRAYS 38
    6.2 DIRAC-LIKE FIELD EQUATION 38
    6.3 DERIVATIVE TERMS 39
    6.4 SU(4) INTERACTION TERM 40
    6.5 DIMENSION ARRAY QUADPLEX SU(4) S MATRICES 41
    6.6 ELECTROWEAK SU(2)⊗U(1) INTERACTION TERMS 41
    6.7 MASS TERMS 44

6.8 GENERAL FORM OF THE STRONG INTERACTION LAGRANGIAN 44

6.9 FORM OF THE ELECTROWEAK INTERACTION LAGRANGIAN TERMS 44

6.10 COSMOS THEORY DIMENSION ARRAYS DOVETAIL WITH QUADPLEX WAVE FUNCTIONS 45

7. GENERATION, LAYER AND CONNECTION GROUPS MATRICES 53

7.1 THE NUMBER OF GROUPS IN THE UST OF OUR UNIVERSE 54

7.2 THE SPECIES COLUMNS OF THE UST 55

7.3 THE REPRESENTATIONS OF GENERATION, LAYER AND CONNECTION GROUPS IN THE UST 56

    7.3.1 Fundamental Generation Representations 56

    7.3.2 Layer Representations 56

    7.3.3 Connection Group Representations 56

    7.3.4 ElectroWeak and Strong Groups Representations 57

7.4 CONCLUSION 57

APPENDIX 7-A. GENERATION AND LAYER 63

7-A.1 U(4) GENERATION GROUPS 63

7-A.2 U(4) LAYER GROUPS 64

APPENDIX 7-B. CONNECTION GROUP SYMMETRIES AND HYPERCOSMOS SPACE-TIME COORDINATES 67

7-B.1 HYPERCOMPLEX COORDINATES TRANSFORMED TO SYMMETRY GROUPS IN OUR UNIVERSE N = 7

.... 67

    7-B.1.1 The U(2) Connection Groups 70

        7-B.1.1.1 Horizontal Lines 71

        7-B.1.1.2 Vertical Lines 71

    7-B.1.2 The Connection Groups are UltraWeak Interactions 72

7-B.2 MEGAVERSE WITH SIX REAL SPACE-TIME COORDINATES (DIMENSIONS) 72

7-B.3 MAXIVERSE WITH EIGHT REAL COORDINATES (DIMENSIONS) 74

7-B.4 DETERMINING THE CONNECTION GROUPS FOR A SPACE 74

8. PERIODIC TABLE OF UST FERMIONS 77

8.1 FERMION DIRAC EQUATION 77

8.2 GENERAL FORM OF WAVE FUNCTION 78

8.3 SU(4) INTERACTIONS AND REPRESENTATIONS 79

8.4 SU(2)⊗U(1) ELECTROWEAK GROUPS 81

8.5 GENERATION GROUPS 83

8.6 LAYER GROUPS 85

8.7 CONNECTION GROUPS 87

    8.7.1 Vertical Connection Group Representations 87

    8.7.2 "Horizontal" Connection Group Representations 87

8.8 QUADPLEX ENHANCEMENT OF THE UST 87

    8.8.1 Quadplex Structuring and UST Layers 88

    8.8.2 HyperCosmos Spaces of the Second Kind 88

8.9 THE INTRICATE INTERACTIONS OF THE UST 88

9. UNIFICATION AND FUNDAMENTAL REFERENCE FRAMES (FRF) 91

9.1 UNIFICATION IN OUR UNIVERSE 91

9.2 THE FRF FOR COSMOS SPACES 91

9.3 THE FUNDAMENTAL REFERENCE FRAME (FRF) AS A "REST FRAME" FOR REFERENCE FRAMES 91

    9.3.1 HyperUnification Space 92

9.4 CONTENTS OF EACH FRF IN 4 DIMENSIONS 95

9.4.1 Rotations Induced by GR Transformations 95
9.4.1 Rotations Induced by GR and Internal Symmetry Transformations 95
9.5 REPLICATES GENERATED BY TRANSFORMATIONS IN HYPERCOSMOS SPACES 99
9.6 PORTRAIT OF A HYPERCOSMOS SPACE AND ITS ASSOCIATED HYPERUNIFICATION SPACE 102
9.7 REDUCTION OF AN FRF'S CONTENTS TO ONE NON-ZERO DIMENSION IN A HYPERUNIFICATION
SPACE 103
9.8 FRFS AND REPLICATES 104

10. THE FULL HYPERUNIFICATION SPACE 105
   10.1 GENERAL RELATIVISTIC TRANSFORMATION 106
   10.2 FULL HYPERUNIFICATION SPACE FRF 106

11. THE HYPERCOSMOS SPACES OF THE SECOND KIND 111
   11.1 TWO HYPERCOSMOSES 112
   11.2 TYPES OF TRANSFORMATIONS 112
   11.3 SECOND KIND HYPERUNIFICATION SPACES 113

12. THE ULTRAUNIFICATION (UU) SPACE OF THE FULL HYPERUNIFICATION SPACE 117
   12.1 RELATION BETWEEN THE FOUR LEVELS 118
   12.1.1 A Qualitative View of Cosmos Spaces *119*
   12.2 REDUCTION OF THE ULTRAUNIFICATION SPACE FRF TO ONE DIMENSION 119

13. FROM DIMENSIONS TO MASS-ENERGY 121
   13.1 GENERATION OF DIMENSIONS WITH MASS-ENERGY FROM GENERAL RELATIVISTIC
   TRANSFORMATIONS 121
   13.2 GENERATION OF *PARTICLES* WITH MASS-ENERGY FROM GENERAL RELATIVISTIC
   TRANSFORMATIONS 121

14. COSMOS THEORY AND THE UST 123
   14.1 UNIFYING PRINCIPLE FOR THE SEQUENCES 123

REFERENCES 125

INDEX 133

ABOUT THE AUTHOR 137

FIGURES and TABLES

Figure 2.1. Map of each universe to its individual corresponding space. Universes may contain sub-universes.. 7

Figure 2.2. The dimension array for our 4D universe. It contains 256 valueless dimensions – "place holders." 7

Figure 3.1. The HyperCosmos space spectrum. See Blaha (2022c)............................. 13

Figure 3.2. The HyperCosmos of the Second Kind space spectrum augmented with N = 10 and N = 11 lines. (Spaces with negative space-times may have universes.) See Blaha (2023d).. 13

Figure 3.3. The Cosmos Theory space spectrum Limos spaces for Gambol Theory.. See Blaha (2023e)... 14

Figure 3.4. Diagram of the four levels of spaces of Cosmos Theory. They contain 42 spaces. From Blaha (2023a). 15

Figure 3.5. Four SU(4) Strong interaction groups in the Normal sector and four SU(4) groups in the Dark sector of the UST. Interactions are between any quark of any generation within each layer in the Normal sector and also in the Dark sector.. There is a different SU(4) for each layer in the Normal and Dark sectors totally to 8 SU(4)'s. This diagram appears in Blaha (2023d) and our earlier books...................................... 16

Figure 3.6. Normal and Dark symmetry groups of UST. SL(2, C) represents the Lorentz group $SO^+(1,3)$. This diagram appears in Blaha (2024i) and our earlier books such as Blaha (2020d). 17

Figure 3.7 Dark and Normal fermions. Unshaded parts are the known fermions. There are 256 fundamental fermions. See Blaha (2018e).. 18

Figure 4.1 The Physical Parent Cosmos universe implied by the consistency condition. The dimension r specifies the space of the universe. It also is the space-time dimension of the universe that is allocated from within the universe space's dimension array. 23

Figure 4.2 The 10 HyperCosmos "physical" spaces that support child universes. The dimension r specifies the space of a universe. It *also* is the space-time dimension of the universe that is allocated from within the universe space's dimension array................. 23

Figure 5.1. Coupling Constants table from Blaha (2024h) where e = 2.718.................. 27

Figure 5.2. Fermion masses based on α/β expressed using powers of 2, π and the base of natural logarithms e. From Blaha (2024i)... 30

Figure 5.3. Sequence 2 fractionation of a particle into 32 gambols. Each gambol is a copy of $(2^5)^{-1}$ times the mass and structure (but not the spin or internal quantum numbers of the particle. The gambol acquires these aspects by inheritance from the particle. Each gambol acts like the particle except for gambol mass and gambol temperature. 31

Figure 5.4. Sequence 1 fractionation of a particle into 32π gambols. Each gambol is a copy of $(\pi 2^5)^{-1}$ times the mass and structure (but not the spin or internal quantum numbers) of the particle. The gambol acquires these aspects by inheritance from the particle. Each gambol acts like the particle except for gambol mass and gambol temperature. 32

Figure 6.1. The dimension array for the γ matrices of the derivative terms. There is one subarray for each space-time index of the $r = 4$ UST of our universe. The y, z, u, and v indices indicate the four Quadplex coordinate systems. Each of the 16 γ submatrices is a 4×4 Dirac γ matrix. The $^y\gamma^\mu$ matrix is for the coordinate system of our universe with $\mu = 0, 1, 2,$ and 3. 46

Figure 6.2. The dimension array for the 16 U(4) T_k matrices of the SU(4) interaction terms of the $r = 4$ UST of our universe. Each of 15 submatrices is a 4×4 SU(4) matrix. The 16th submatrix is the identity matrix. These matrices are for the first UST layer. Other UST layers have different SU(4) groups. The submatrices of these groups have the same form as the above submatrices. The dimension array supports a map to the set of U(4) matrices. 47

Figure 6.3. There are four pairs of fermions for the eight fermions in the four UST generations of the first layer as one can see from the sequences in eq. 6.2. Each column corresponds to a representation of SU(2)⊗U(1) denoted with an upper index. The four entries in each column are its SU(2)⊗U(1) 4×4 submatrices denoted with lower indices in its real-valued SU(2)⊗U(1) representation. The other UST layers have different SU(2)⊗U(1) groups and representations. The submatrices of these groups have the same form as the above submatrices. Dimension arrays support maps to the set of U(4) matrices in any Cosmos space. 48

Figure 6.4. The wave functions of the eight first generation, first layer UST fundamental fermions for use in the Strong Lagrangian ordered by their sequences and positions within the sequences. The other wave functions of the other generations and layers are similar in form. The wave functions are denoted by the fermion's acronym. 49

Figure 6.5. Eight dimension representation of the pair of four dimension SU(4) representations of the fermion sequences in generation 1, layer 1 of Normal fermions. 49

Figure 6.6. Fermion wave functions terms ordered as four sets of fermion wave function pairs for ElectroWeak use use in Fig. 6.7. 50

Figure 6.7. Four ElectroWeak representations for one generation of one layer. The representations correspond with the order of the fermions in the fermion wave function vector of Fig. 6.6. 50

Figure 6.8. Diagonal Lagrangian fermion mass matrix for generation 1, layer 1 of the Strong SU(4). Masses are denoted by fermion symbol. 50

Figure 6.9. Diagonal Lagrangian fermion mass matrix for generation 1, layer 1 of the ElectroWeak Lagrangian. Masses denoted by fermion symbol. Other components are zeroes (not shown). The dimension array supports a map to this set of ElectroWeak mass matrices.. 51

Figure 7.1. The transformed/broken sets of symmetries in the UST. The darkened parts have not as yet been found experimentally. The one undarkened line is for experimentally known groups. Note each item above has an 8 real dimension representation. Note the seven U(2) Connection groups. The SL(2, **C**) representation has four coordinates... 57

Figure 7.2. The dimension array for the 16 U(4) T_k matrices of U(4) and SU(4) interaction terms. Each of 16 submatrices is a 4 × 4 U(4) matrix. The 16th submatrix is the identity matrix... 58

Figure 7.3. The UST Fermion particle spectrum and partial examples of the pattern of fermion mixing of the Generation groups and of the Layer groups. Unshaded fermion dots are the known fermions, The lines on the right side show Layer group mixing (for Normal and Dark matter) with the mixing among all four layers for each of the four generations individually. There are four Layer groups for Normal matter and four Layer groups for Dark matter. There are 256 fundamental fermions. From Blaha (2018e)..... 59

Figure 7.4. Symbolic display of the 8 Generation group representations for fermions of each fermion species within each layer. The "do" superscript indicates "down" type. . 60

Figure 7.5 Symbolic display of Layer Group representations. Each species has separate representations for each generation. Unshaded parts are the known fermions including μ, $ν_μ$, τ, and $ν_τ$ leptons. Shaded parts are yet to be found. From Blaha (2018e). 61

Figure 7.6. ElectroWeak SU(2)⊗U(1) form. Together with SU(4) (not shown) they span an entire layer. 62

Figure 7-B.1. The four UST layers internal symmetry groups (and space-time) with SU(4) before breakdown to SU(3)⊗U(1). Note the left column of blocks are combined below to specify a 4 dimension real space-time plus seven U(2) Connection groups. Note each layer has 64 dimensions = 56 + 8 dimensions. 69

Figure 7-B.2. The seven U(2) Connection groups (shown as 10 lines) between the eight UST blocks. Connection groups are obtained by transfering 28 dimensions from UST space-time to internal symmetries with the consequent reduction of the space-time from four octonion (complex quaternion) coordinates to four real coordinates. The Connection groups generate rotations and interactions between corresponding fermions and vector bosons of each pair of blocks. The Normal and Dark sector U(2) vertical connections above (E, F, G) represent the same U(2) groups. 70

Figure 7-B.3. MEGAVERSE has four UST copies. An SU(4) internal symmetry Connection group maps between corresponding fermions in the four copies: fermion by fermion. An additional U(1) Connection group applies to every corresponding fermion. It is not shown in this figure. 73

Figure 7-B.4. The SU(4) Connection Group of MEGAVERSE connecting fermions in the four UST "copies" blocks. An additional U(1) Connection group applies to every corresponding fermion. It is not shown in this figure.. 73

Figure 7-B.5. The SU(8) Connection Group of Maxiverse connecting fermions in the four MEGAVERSE "copies" blocks. 74

Figure 8.1. Pairs of SU(4) representations for the two sequences of fermions appear in each of the four layers. Each layer has a different SU(4) interaction group. 80

Figure 6.7. Four ElectroWeak representations for one generation of one layer. The representations correspond with the order of the fermions in the fermion wave function vector of Fig. 6.6.. 81

Figure 6.6. Fermion wave functions terms ordered as four sets of fermion ElectroWeak representations using pairs of fermions in Fig. 6.7... 81

Figure 8.2. Four layers of ElectroWeak representations. Each layer has a set of sixteen ElectroWeak representations. Each ■ corresponds to a one generation, one layer set of four ElectroWeak representations as in Fig. 6.7 above. This figure has 64 Normal ElectroWeak representations in total. Their total number of dimensions is 128 = 2×64 – corresponding to the size of the UST Normal sector. (The complete specification has a size of 256 if one takes account of the Dark sector as expected in the UST.) Note each layer has a different ElectroWeak group and interaction, and 16 representations. 82

Figure 8.3. Eight Generation group representations for each layer. We specify each four dimension Generation representation with the ■ symbol. The representations are displayed in block diagonal form. In each layer the eight fermion Generation representations are for the species. Each layer has a different Generation group. 84

Figure 8.4. Eight species Layer group representations for each generation. We specify each four dimension Layer group representation with the ■ symbol. The representations in the figure are in block diagonal form. The number k of the Layer group $A_{kL}{}^{a\mu}(y)$ is chosen to be the number of the generation of its representations................................... 86

Figure 8.5. The transformed symmetries of the UST modified to the Quadplex formulation with SL(2, C) for the y, z, u, and v coordinate systems. The Dark sector uses the same coordinate systems since Dark matter/energy occupies well-defined positions in the universe. The darkened parts have not as yet been found experimentally. The one undarkened line is for experimentally known groups. The SL(2, C) representation has four coordinates. 89

Figure 8.6. Second Kind HyperCosmos layers. There is no Dark sector. 90

Figure 9.1. The HyperCosmos spaces spectrum related to a HyperUnification space space-time dimension r′ and its dimension array. Note the spaces with r′ > 18 are outside the HyperCosmos set of 10 spaces. However they are made to have the same form as the 10 HyperCosmos spaces. 94

Figure 9.2. The contents of the ten types of Fundamental Reference Frames for the ten spaces. The elements are fermions or dimensions or symmetry fundamental representation dimensions as the case may be.. 96

Figure 9.3. Possible initial FRF contents of five of the ten types of Fundamental Reference Frames for fermions. The primes distinguish different fermions.................. 97

Figure 9.4. Possible map of initial FRF contents to five of the ten types of Fundamental Reference Frames for internal symmetry dimensions. The groups may undergo a further breakdown after being mapped... 98

Figure 9.5. The 16 fermions that constitute the initial set that is then replicated 16 times to obtain the 256 fermion spectrum for N = 7 space (our universe)............................ 98

Figure 9.6. Visualisation of the T_E transformation replication pattern in the HyperUnification space and its FRF. The FRF has $2^{r/2-2}$ replicates of the basic 16 dimension set. 99

Figure 9.7. The UST 16 × 16 fermion spectrum of our universe tentatively arranged as SU(4)-plets that correspond directly with SU(4) (or SU(3)⊗U(1)) fermions. There are four layers. Each set of 4 fermions has 4 generations matching the number of rows in each layer. This Periodic Table is broken into Normal and Dark sectors. 100

Figure 9.8. UST Fermion particle spectrum and partial examples of the pattern of mass mixing of the Generation groups and of the Layer groups. Unshaded parts are the known fermions, The lines on the left side (only shown for one layer) display the Generation Group mixing within each layer. The Generation mixing occurs within each layer using a separate Generation group for each layer. The lines on the right side show Layer group mixing (for Dark matter) with the mixing among all four layers for each of the four generations individually. There are four Layer groups for Normal matter and four Layer groups for Dark matter. There are 256 fundamental fermions. The UST have the same fermion spectrum. 101

Figure 9.9. The "initial" distribution of sets of N = 7 symmetry groups. Each set is distinct and supports interactions only for the corresponding set of fermions (separately for Normal and Dark fermions). *Thus each set of 16 fermion generations has its own quantum numbers and interactions.* Each U(4)⊗U(4) set has a 16 real-valued dimension representation, which is importance when we consider Fundamental Reference Frames. 102

Figure 9.10. The transformed/broken sets of symmetries in UST and in N = 7, r = 4 HyperCosmos space. Note each element has a 16 real dimension representation. This depiction is also evident in the UST. The SL(2, **C**) representation has four coordinates. 102

Figure 9.11. Diagram of the relation between a HyperCosmos space and its associated HyperUnification space with Fundamental Reference Frames indicated. The HyperUnification space sets the dimension array d_{dN} for the HyperCosmos space using purely HyperUnification General Relativity. Fundamental Reference Frames are like the rest frames of particles. They reduce the number of elements to a core set of elements. The elements are dimensions or fermions or gauge vector symmetry fundamental representation dimensions. 103

Figure 10.1. HyperCosmos FRF transformations in the Full HyperUnification Space. 108

Figure 10.2. Form of a HyperUnification transformation. It is also the form of the dimension array of 42 dimension space-time. Block A is square. The figure is not drawn to scale. 109

Figure 10.3. The r = 42 space-time HyperUnification vector form. 110

Figure 11.1. The HyperCosmos of the Second Kind space spectrum. The space for our universe, is Blaha number N = 7, with Cayley-Dickson number 3 (which corresponds to octonions) is in bold type. Note changed d_d column relative to the HyperCosmos. 113

Figure 11.2. The HyperCosmos of the Second Kind 8 × 16 fermion spectrum tentatively arranged as SU(4)-plets that correspond directly with SU(4) (or SU(3)⊗U(1)) fermions. There are four layers. Each set of 4 fermions has 4 generations matching the number of rows in each layer. This Periodic Table is broken into Normal and Dark sectors. The absence of a Dark sector is indicated by the darkening............ 114

Figure 11.3. The Second Kind "initial" distribution of sets of Blaha number N = 7 symmetry groups. Each set is distinct and supports interactions only for the corresponding set of fermions (separately for Normal and Dark fermions). *Thus each set of 16 fermion generations has its own quantum numbers and interactions.* Each U(4)⊗U(4) set has a 16 real-valued dimension representation, which are of importance when we consider Fundamental Reference Frames. The absence of a Dark sector as indicated by the darkened part. 114

Figure 11.4. The transformed/broken sets of symmetries in the Blaha number N = 7, r = 4 Second Kind HyperCosmos space. Note each element has a 16 real dimension representation. This depiction is also evident in a Second Kind UST. The SL(2, **C**) representation has four coordinates. The absence of a Dark sector as indicated by the darkened part. 115

Figure 11.5. The three U(2) Connection groups (shown as 3 lines) between the eight UST blocks in the Blaha number N = 7 Second Kind HyperCosmos. The Darkened part is not present in the Second Kind case. Connection groups are obtained by transfering 12 dimensions from the UST space-time to internal symmetries with the consequent reduction of the space-time from four octonion (complex quaternion) coordinates to four real coordinates. The Connection groups generate rotations and interactions between corresponding fermions and vector bosons of each pair of blocks. This figure is abstracted from the corresponding figure in Appendix 7-B. 116

Figure 12.1. Diagram of the four levels of the Cosmos. They consist of 42 spaces. ... 120

Introduction

This book continues the presentation of Cosmos Theory in *The Dawn of Dynamic Cosmos Dimension Arrays.* Starting from a discussion of the structure and basis of Cosmos Theory in tensors, spin and fractals, it proceeds to develop the dynamics of fermions embodied in the Quadplex Dirac equation. This analysis describes the form and content of the ElectroWeak, Strong, Generation Group, Layer Group, and Connection group interactions.

The dimension array for a four dimension universe such as ours is shown to describe the Dirac matrices, ElectroWeak generator matrices, and the generator matrices of the other interaction groups: the Generation, Layer and Connection Groups. The Dirac equation for Cosmos Theory fermions is explored in detail using a Quadplex formalism. Mass terms and coupling constants seen in Nature are now numerically determined and are in general agreement with experiment. The book shows the Quadplex formalism is well-adapted to Cosmos Theory dynamics studies.

The Generation, Layer and Connection groups are explained in detail. The number of fundamental representations for all interaction groups in the Dirac equation is 592 representations, which shows the depth of the dynamic Dirac equation analysis.

The unification of Cosmos Theory spaces and groups occurs on four levels: 20 HyperCosmos and Second Kind HyperCosmos spaces, 20 Unification spaces, 1 Full Unification space, and 1 UltraUnification space. These spaces are described in detail based on Fundamental Reference Frames (FRF) which has the ability to build a space of many dimensions from one dimension. FRF's are the "seeds" of spaces.

The 88 dimension space can be used to build the entire Cosmos spectrum of spaces from one dimension. The qualitative features of these spaces accords well with Parmenides view of the Cosmos: unchanging and timeless. Cosmos spaces are beyond time. Matter and energy change but the Cosmos spectrum of spaces does not change.

All in all, Cosmos Theory spaces are unchanging conceptual entities without beginning or end. Universes may be viewed as generated from a single particle, in accord with a Consistency Condition, in the 88 dimension space FRF according to the definition of Cosmos spaces. The simplicity of these concepts recommends themselves. See chapter 13.

One of the results of these investigations is a Periodic Table of Fundamental Fermions whose structure compares qualitatively with the Chemical Periodic Table of Elements.

This book, and the author's previous books, shows the Cosmos is a compact, highly structured set of spaces describing universes and particles.

1. Space and Spin

A space[1] has two defining characteristics: a space-time that will be used for dynamical evolution, and a set of spins that determine other aspects of dynamical evolution. A space-time is characterized by an integer r specifying the number of its dimensions and a metric with a signature distinguishing between time and space dimensions.

In our Cosmos Theory we introduce an array of dimensions, a dimension array, for each space. It specifies the number of space-time dimensions and the number of internal symmetry related dimensions. In earlier books we showed that the dimension array of a space arises from the same considerations[2] as the space's γ-matrices, which are intimately related to spin.

Dimensions determine spin. The spin of the fermion of lowest spin determines the size and shape of a space's dimension array.

The spin of a space is directly related to the dimension of its space-time. In the case of our universe's space-time the minimal spin is ½—the spin of fundamental fermions. Other universes and higher dimension universes have minimal fermion spins that are odd integer multiples of ½.

---

[1] We distinguish between a space and a universe in chapter 2. A space is a specification of a structure consisting of dimensions that have no initial values. After defining/creating a universe the dimensions may be mapped to symmetry groups, fermions and so on. Some dimensions are used for space-time. The other dimensions are used for internal symmetries.

[2] Chapter 4.

2. Structure and Substance

Cosmos Theory distinguishes between spaces and universes. Spaces are abstract structure specifications of arrays of valueless dimensions. They are without substance. Each space in the spectrum of Cosmos spaces has an array called a dimension array.

2.1 Structure – Dimension Arrays

A dimension array specifies a structure. It contains a set of dimensions that have no defined value. Dimension arrays acquire value by being assigned to specific features of a universe. Thus, for a universe, there is a dimension array map to space-time dimensions and internal symmetry group fundamental representation dimensions. There is also a dimension array map to the spectrum of fundamental fermions. And there are also dimension array maps to the set of γ-matrices, the set of Strong Interaction SU(4) generator matrices, and ElectroWeak generator matrices. And so on.

Dimension arrays specify structure that maps to features of universes and their dynamics.

2.2 Substance - Universes

Universes are entities that have substance – matter and energy. Each universe also has a corresponding HyperCosmos space square dimension array[3] that specifies a set of dimensions for each of its internal symmetry fundamental representations, and also for the space-time dimensions of the universe. See Fig. 2.1.

2.3 Separation of Structure from Substance

We separate the concepts of space and universe for several reasons. It provides structure in a manner independent of details of universes. Many models of our universe tend to assume the number of space-time dimensions and the set of internal symmetries. Thus they generally give form to a model without specifying the origin of these aspects of the universe.

By separating structure from substance we can form a general specification of universes of any dimension. This generality enables us to treat universes uniformly and to avoid ersatz universes that are not Physically acceptable.

We will see later that dimension arrays specify many aspects of a universe and its dynamics such as the form of internal symmetry generator matrices. Thus the dimension array of a universe compactly specifies a host of its features.

The separation into structure and substance enables us to consider multiple universes having the same space specification (dimension array) on a uniform basis. One then has a general theory of universe structure.

---

[3] Cosmos Theory also has a set of spaces – HyperCosmos Spaces of the Second Kind that are not square. These space's dimension arrays are rectangular with dimension array width and length in the ratio 2.

A separation of the theory of the Cosmos into a space part (without substance) and a set of universes gives a clear picture of the more general case of a Cosmos with many universes of many different space (dimension array) types.

2.4 Space = Dimension Array

Although one may view a Cosmos space as having a number of parts we view a space as only specifying a dimension array. A dimension array is a two dimension array of dimensions. Dimensions do not have a numerical value in themselves. Dimensions may be viewed as "placeholders." The set of dimensions in a dimension arrays acts as a template for maps to features of universes such as fundamental fermion spectrums and internal symmetry groups.

Fig. 2.2 shows a dimension array for our 4D universe with each "dot" representing a dimension. Because of our specification of dimensions as primitives without value we view a dimension as a primitive term from a Logic viewpoint. A dimension array is an array of primitives. A space is an identifier for dimension arrays. Spaces may be viewed as analogues of other things as we do later in discussions of asymmetric tensors and fractal curves. The set of spaces and their dimension arrays may be viewed as generated from dynamic equations as we did in our books describing a proposed ProtoCosmos model. Yet they remain arrays of valueless dimensions. See Fig. 2.2.

2.5 How does the Overall Structure of Cosmos Theory Arise?

Since the structure of the Cosmos spectrum of spaces does not have substance, its set of spaces may be viewed as a set of dimension arrays. The spectrum of spaces is a conceptual definition without a need for the specification of an origin. In that respect it is analogous to Plato's Theory of Ideals. The set of Cosmos spaces is timeless and unchangeable.

The universes that implement a space definition appear to be created entities. Later we will see that an initial parent universe(s) must exist that contains (possibly chains of) sub-universes. The origin of a parent universe is problematic. We can say, as we do later, that the parent universe(s) must satisfy an energy consistency condition for it to exist. Beyond that, we can only view its origin as beyond our current endeavors. The dynamics that generates/creates sub-universes also remains to be investigated.

We view the "creation" of the Parent universe as the beginning of a *time*. (There can be no time (or dynamics) prior to the existence of a space-time. A space-time requires existence within a universe. Thus time begins within a parent universe at its creation point.) The times of the sub-universes, including their creation times, are relative to the parent universe time. "No time before time."

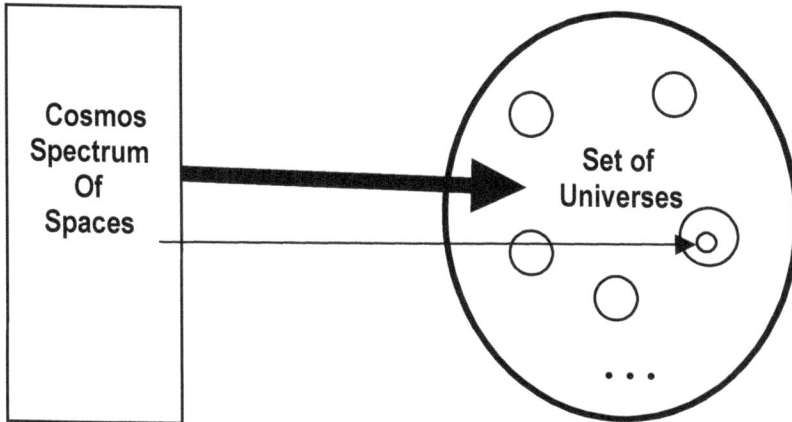

Figure 2.1. Map of each universe to its individual corresponding space. Universes may contain sub-universes.

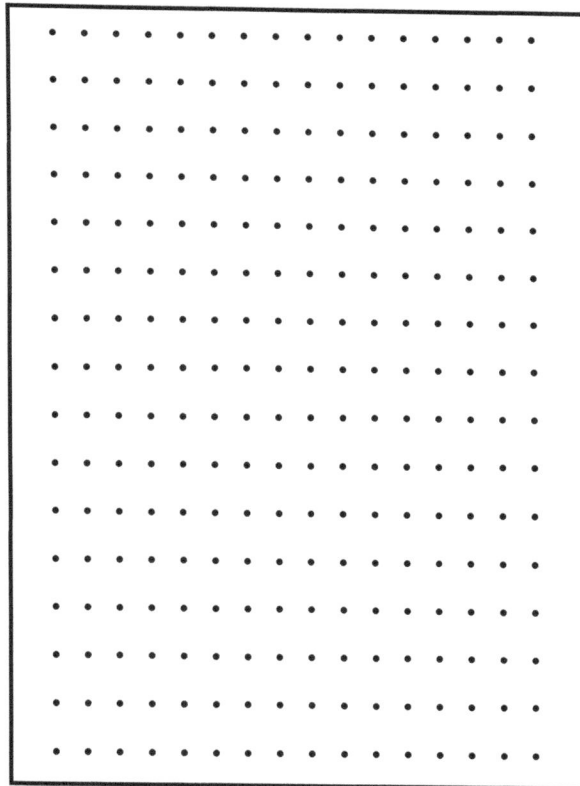

Figure 2.2. The dimension array for our 4D universe. It contains 256 valueless dimensions – "place holders."

3. Overall Structure of Cosmos Theory

This chapter outlines some of the features of Cosmos Theory. The Fundamental Reference Frame and the 42 and 88 dimension unification spaces are discussed later. The reader is referred to our recent books on these topics.

3.1 Cosmos Theory Spaces

There are 10 physical HyperCosmos spaces.[4] They are listed in Fig. 3.1. These spaces are numbered with even dimensions r, denoted $r = 0, 2, 4, \ldots, 18$. They each define a square dimension array with 2^{r+4} components. In each space the dimension array has r space-time dimensions and $2^{r+4} - r$ dimensions for internal symmetries. The internal symmetries found throughout the spaces are: SU(4) (or broken to SU(3)⊗U(1)), SU(2)⊗U(1), U(1), and a space-time symmetry. Figs. 3.5 and 3.6 display the fundamental fermion spectrum and internal symmetries implied by the dimension array of the Unified SuperStandard Theory (UST) of our universe.

In addition to these ten spaces there is a set of spaces called Limos (Fig. 3.3) that implement Gambol Theory within Cosmos Theory. See Blaha (2023e).

The number of physical HyperCosmos spaces, ten, is set by a consistency condition[5] that balances the internal thermodynamic expansion pressure of a universe against the external constraining vacuum Casimir force. This condition eliminates the need for a dynamics before the creation of universes. Time does not exist before the existence of universes. There is nothing before the beginning except structure without substance.

There is also a set of spaces that we call HyperCosmos Spaces of the Second Kind that are similar to HyperCosmos spaces. See Fig. 3.2. They are discussed in previous books by the author and in chapter 11.

There are additional spaces that implement the unification of interactions. These spaces are described later. The 42 and 88 dimension unification spaces complete the list of Cosmos Theory spaces. Fig. 3.4 lists the set of Cosmos spaces (except for Limos.)

Dimension arrays grow in size from HyperCosmos space by space by factors of 4. See the d_{dN} column in Fig. 3.1. Fig. 3.5 shows the Unified SuperStandard Theory (UST) dimension array of our universe populated by fundamental fermions. Fig. 3.6 shows the UST dimension array populated by symmetries.

---

[4] There are also ten HyperCosmos spaces of the Second Kind described in Blaha (2023d) and chapter 11. At present we view these spaces as not Physically needed for our treatment of the Unified SuperStandard Theory (UST) that we developed as a generalization of The Standard Model due to the apparent presence of a Dark sector.

[5] Chapter 4.

We have found that other quantities in the Unified SuperStandard theory (UST), which has an r = 4 HyperCosmos space, also grow by various multiplicative factors: coupling constants, and first generation fermion masses in two sequences.

Later we will see that γ-matrices, SU (4) matrices, and SU (2) ⊗U (1) matrices of Dirac fermion dynamic equations have a format similar to dimension arrays in the UST. *Thus dimension arrays are integral to the UST and Hyper Cosmos universes in general.*

This short summary of previous work by the author does not present many important details. Its purpose is to give an initial view prefatory to the later chapters.

3.2 The Cause of Dark Matter/Energy in Cosmos Theory and UST

Dark matter/energy is known to exist in our universe due to its gravitational effects. It is called Dark because no radiation is detected from it. The reason for its Darkness and its nature is not fully known.

Our Cosmos Theory, and the UST in particular, provide a direct explanation. In the UST the four layers of Normal each have a separate SU(4) (or SU(3)⊗U(1)) Strong Interaction group and a separate SU(2)⊗U(1) ElectroWeak group. Similarly under the UST (and Cosmos Theory) we have sets of SU(4) and SU(2)⊗U(1) groups for each of the four Dark layers. See Fig. 3.6. These Dark symmetry groups are different from the symmetry groups of Normal matter although they have similar defining features. They do not have "cross terms" with Normal matter and energy (except possibly through Connection group interactions.)

3.2.1 Visibility of Dark Matter and Energy

Thus the Dark ElectroWeak interactions, and the photons of Electromagnetism in particular, are not "seen" by normal matter detectors. That is the essential reason for the Darkness in the UST formulation for our universe.

The three additional "Normal" layers are also Dark experimentally because their ElectroWeak interactions are not the same as those of the first layer. We can't see these layers using first layer photons. Thus we have the sector labeled Dark and three Normal layers that are also Dark from our first layer photon view. See Fig. 3.7.

3.2.2 Dark ElectroWeak Gauge Theory

Dark matter and energy are not visible because their possible interactions with Normal matter (observation tools) are prevented due to their implementation with *independent* sets of ElectroWeak gauge fields.[6] See Fig. 3.6. Dark photons cannot interact with Normal matter. Thus the Darkness. Normal photons cannot interact with Dark matter. We see this feature in experiment where we cannot manipulate Dark matter electromagnetically. *These features are implicit in the formulation of the UST.*

We take the nature (structure) of Dark matter and energy to be the same as Normal matter and energy in the UST.

---

[6] We augment this discussion below with the introduction of Connection groups and interactions. These interactions connect Normal and Dark matter and energy.

The general features and nature of Normal and Dark matter and energy apply to higher space-time dimension Cosmos universes although on a larger scale.

3.2.3 Interpenetrability of Normal and Dark Matter

Normal matter has substantiality. Normal matter chunks cannot occupy the same space. The cause of substantiality in Normal matter is the ElectroWeak forces. If there is no first layer ElectroWeak force felt by Dark matter, then ElectroWeak substantiality does not exist for Normal interspersed with Dark matter.

The forces within Dark matter are Dark forces based on Dark ElectroWeak interactions. These Dark forces do not affect Normal matter and energy. Similarly Normal ElectroWeak forces do not affect Dark matter and energy.

Normal matter is interpenetrable with Dark matter. Normal matter and Dark matter may occupy the same spatial region.

The above subsections show that Cosmos Theory and the UST account for the known general features of Dark matter and energy by having different ElectroWeak and Strong Interaction groups in the Dark sector and in the second, third and fourth layers of the Normal sector. The first layer of the Normal sector has the only ElectroWeak and Strong Interaction groups with which we are familiar. The masses and energies of particles and interactions of this layer are the only ones that are familiar from experiment. The rest is Datkness.

3.2.4 Connection Group and Interactions between Normal and Dark Matter

The UST posits the existence of a set of seven Connection SU(2) groups and interactions that connect Normal and Dark matter and energy.[7] These interactions must be extremely weak since Dark matter and energy has not been seen experimentally. We will call the groups and interactions *UltraWeak*. Experiments seeking to "find" Dark matter and energy would necessarily rely on UltraWeak interactions. UltraWeak masses and energies remain to be found. If found they would modestly modify the results of the above subsections. See Chapter 7 and its appendices for discussions of Connection groups.

3.3 The Relative Distribution of the Normal and the Dark in Our Universe

We now turn to the analysis of the distribution of Normal and Dark matter and energy based on an Equipartition Principle for Fermions and Gauge Fields. We will suggest[8] a rationale for the dominance of Dark mass-energy:

3.3.1 Equipartition Principle for Particle Degrees of Freedom

In a closed system at equilibrium the thermal energy of a system is equally partitioned (distributed) among its degrees of freedom. The Equipartition Principle is well known. The application of this principle to the beginning of the universe *when all particles were massless* and all symmetries are unbroken suggests that the distribution of mass-energy should be the same for all degrees of freedom at that

---

[7] See Appendix 7-B.
[8] The material in this section appeared in Blaha (2016a) and (2018e).

time. Thus there should be approximately equal energies for the UST 256 fermions and 224 vector bosons.

　　　We now estimate the relative proportion of Normal and Dark matter in the universe at its beginning based on this Equipartition Principle.

3.3.2 Proportion of Dark Mass-Energy in the Universe

　　　First we note that in the UST there are $4 \times 8 = 32$ first layer fundamental fermions[9] that are visible *with our first layer ElectroWeak theory photons*.

　　　In total, there are 128 Dark sector fermions plus $3 \times 32 = 96$ Dark fermions in the other three layers of the Normal sector totaling to 224. See Fig. 3.7.

　　　There are 256 fermions in the UST spectrum.

　　　Thus 224 of the 256 fermions are Dark yielding a *percentage of Dark Matter equal to 224/256 = 87.5%*.

　　　Recent studies of the proportion of Dark Matter in the universe have yielded an estimate 85% Dark matter.

　　　Thus our estimate based on our fermion Equipartition Principle is reasonable. Two possibilities emerge with respect to the present proportion of Dark Matter:

1.　The percentage has not changed from the Beginning and the approximate estimates are slightly off. The lack of change could be due to the extremely small decay rates of the fermions in the higher layers.

2.　The percentage of matter in the upper layers has decreased due to decay and so the current proportion may be somewhat below 87.5%.

3.3.3 Proportion of Dark Mass-Energy in the Universe

　　　We know of 4 ElectroWeak + 8 SU(3) = 12 known vector boson particles of the 480 UST vector bosons, and of 12 known fundamental fermions[10] totaling 24 in the UST. Thus 456 out of 480 UST particles[11] are Dark yielding a Dark mass-energy of 95% of the universe's mass-energy at the beginning of the universe.

　　　The Dark energy in the universe currently has been estimated to be 68% of the total energy and the energy of Dark Matter is estimated to be 26%. The total is approximately 95% - a value equal to our above approximate estimate of 95%.

---

[9] The first layer has ν, e, u, d, s, c, b, t type fermions in each of 4 generations.
[10] The known fermions are ν, e, u, d, s, c, b, t, μ, $ν_μ$, τ, $ν_τ$ fermions.
[11] 480 = 4 SL(2,C) Coordinates + 4 SU(2)⊗U(1) ElectroWeak + 16 SU(4) Strong + 16 U(4) Generation + 16 U(4) Layer + 256 fermions. See Figs. 3.6 and 3.7.

The calculations of this section 3.3 numerically confirm the separation of Normal and Dark sectors, and the four layers of the UST. This UST structure is directly based on Cosmos Theory structure in general.

THE HYPERCOSMOS SPACES SPECTRUM

Blaha Space Number	Cayley-Dickson Number	Cayley Number C_n	Dimension Array column length	Dimension Array Size	Space-time-Dimension	CASe Group $su(2^{r/2}, 2^{r/2})$
$N = o_s$	n	d_c	$d_{cd} = d_{cr}$	$d_{dN} = d_{dr}$	r	CASe
0	10	1024	2048	2048^2	18	su(512,512)
1	9	512	1024	1024^2	16	su(256,256)
2	8	256	512	512^2	14	su(128,128)
3	7	128	256	256^2	12	su(64,64)
4	6	64	128	128^2	10	su(32,32)
5	5	32	64	64^2	8	su(16,16)
6	4	16	32	32^2	6	su(8,8)
7	3	8	16	16^2	4	su(4,4)
8	2	4	8	8^2	2	su(2,2)
9	1	2	4	4^2	0	su(1,1)
10	0	1	2	2^2	-2	

Figure 3.1. The HyperCosmos space spectrum. See Blaha (2022c).

HYPERCOSMOS OF THE SECOND KIND SPACES SPECTRUM

Blaha Space Number	Cayley-Dickson Number	Cayley Number	Dimension Array size	Space-time-Dimension	CASe Group $su(2^{r/2}, 2^{r/2})$
$N = O_s$	n	d_c	d_{dN2}	r	CASe
0	10	1024	1024×2048	18	su(512,512)
1	9	512	512×1024	16	su(256,256)
2	8	256	256×512	14	su(128,128)
3	7	128	128×256	12	su(64,64)
4	6	64	64×128	10	su(32,32)
5	5	32	32×64	8	su(16,16)
6	4	16	16×32	6	su(8,8)
7	3	8	8×16	4	su(4,4)
8	2	4	4×8	2	su(2,2)
9	1	2	2×4	0	su(1,1)
10	0	1	1×2	-2	
11	-2	½	½	-4	

Figure 3.2. The HyperCosmos of the Second Kind space spectrum augmented with N = 10 and N = 11 lines. (Spaces with negative space-times may have universes.) See Blaha (2023d).

THE Limos SECTOR SPACES SPECTRUM

Blaha Space Number	Cayley-Dickson Number	Cayley Number	Row Size	Dimension Array Size	Space-time-Dimension	CASe Group $su(2^{r/2}, 2^{r/2})$
$N = O_s$	n	d_c	d_{cr}	$d_{dN2} = d_{dr}$	r	CASe
10	0	1	2	2^2	2	U(1)
11	-1	½	1	1^2	4	U(½)
12	-2	¼	½	$½^2$	6	U(¼)
13	-3	1/8	¼	$¼^2$	8	U(1/8)
14	-4	1/16	1/8	$1/8^2$	10	U(1/16)
				⋮		
∞	$-\infty$	0	0	0	$-\infty$	U(0)

Figure 3.3. The Cosmos Theory space spectrum Limos spaces for Gambol Theory.. See Blaha (2023e).

Level

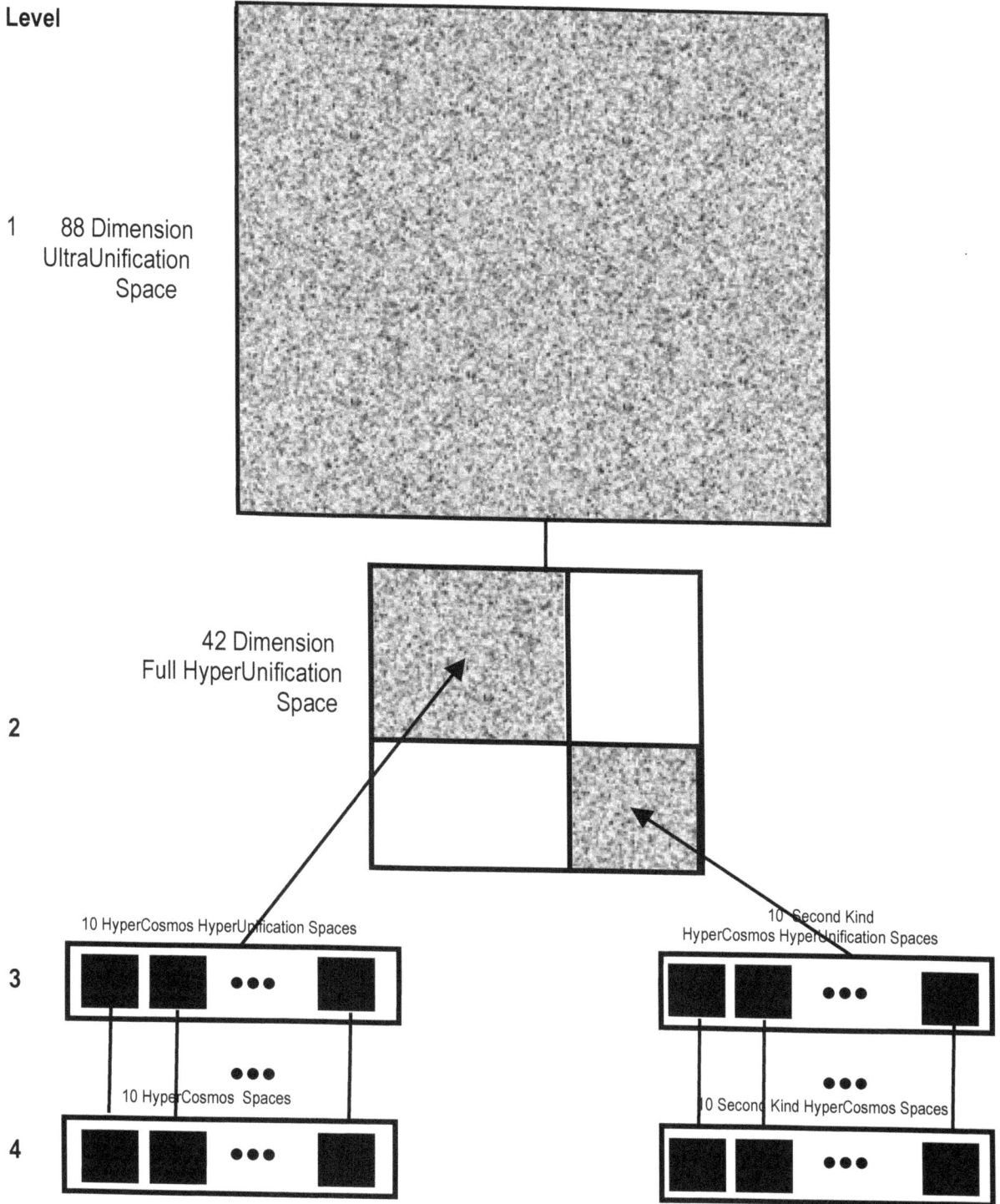

1   88 Dimension
UltraUnification
Space

2   42 Dimension
Full HyperUnification
Space

10 HyperCosmos HyperUnification Spaces

10 Second Kind
HyperCosmos HyperUnification Spaces

3

10 HyperCosmos Spaces

10 Second Kind HyperCosmos Spaces

4

Figure 3.4. Diagram of the four levels of spaces of Cosmos Theory. They contain 42 spaces. From Blaha (2023a).

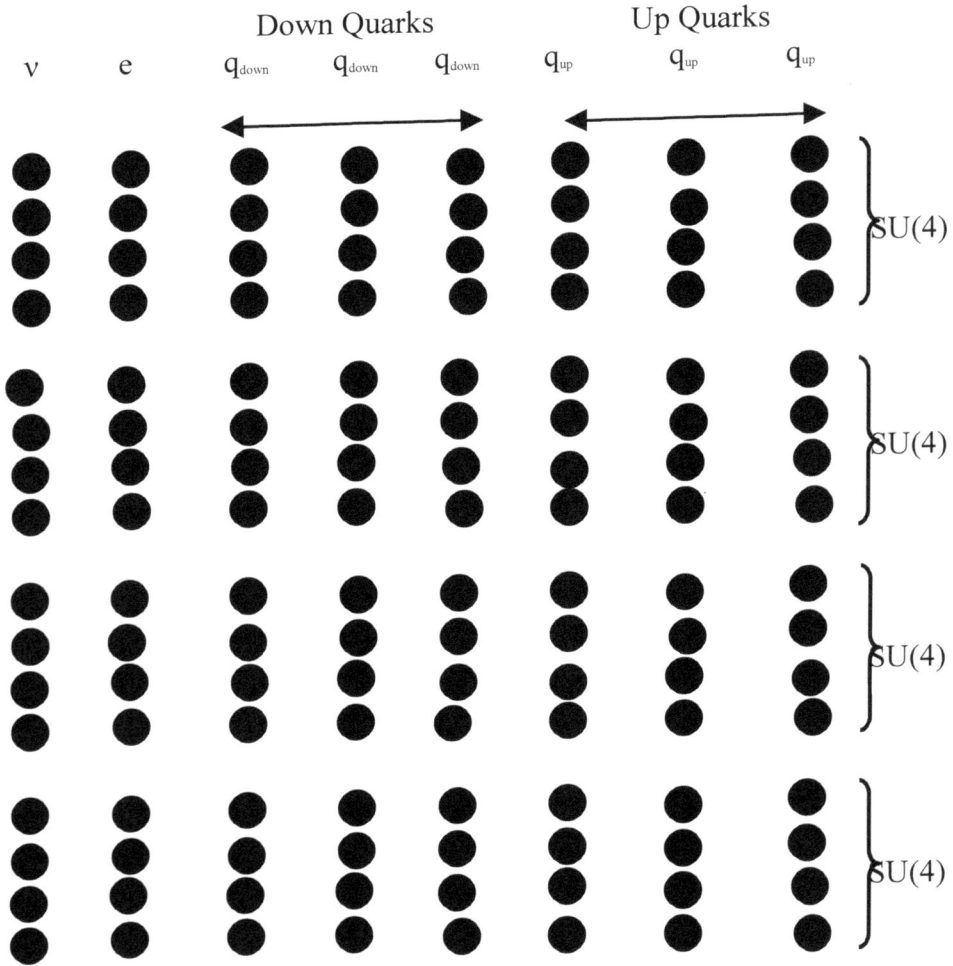

Figure 3.5. Four SU(4) Strong interaction groups in the Normal sector and four SU(4) groups in the Dark sector of the UST. Interactions are between any quark of any generation within each layer in the Normal sector and also in the Dark sector.. There is a different SU(4) for each layer in the Normal and Dark sectors totally to 8 SU(4)'s. This diagram appears in Blaha (2023d) and our earlier books.

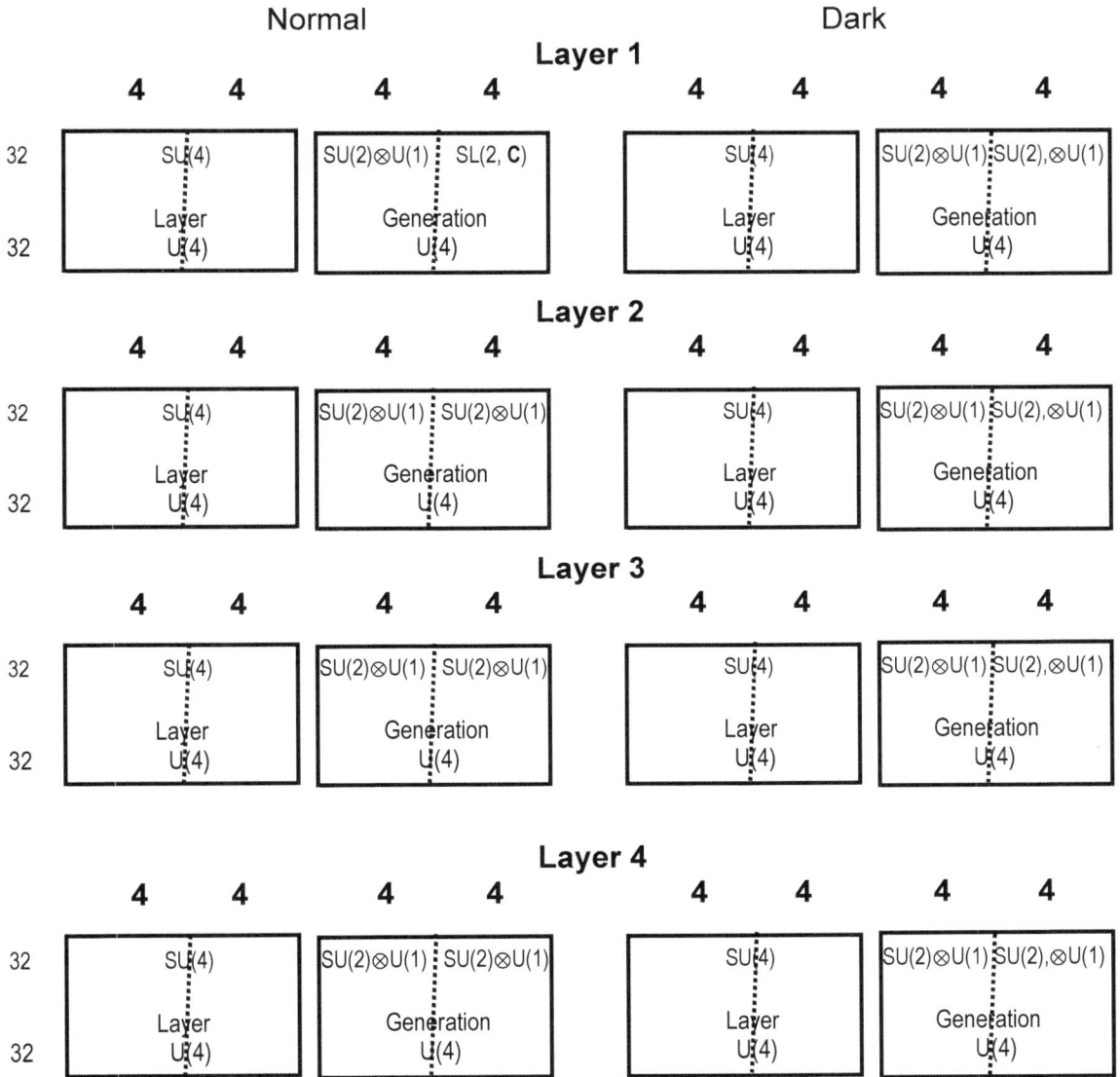

Figure 3.6. Normal and Dark symmetry groups of UST. SL(2, C) represents the Lorentz group SO⁺(1,3). This diagram appears in Blaha (2024i) and our earlier books such as Blaha (2020d).

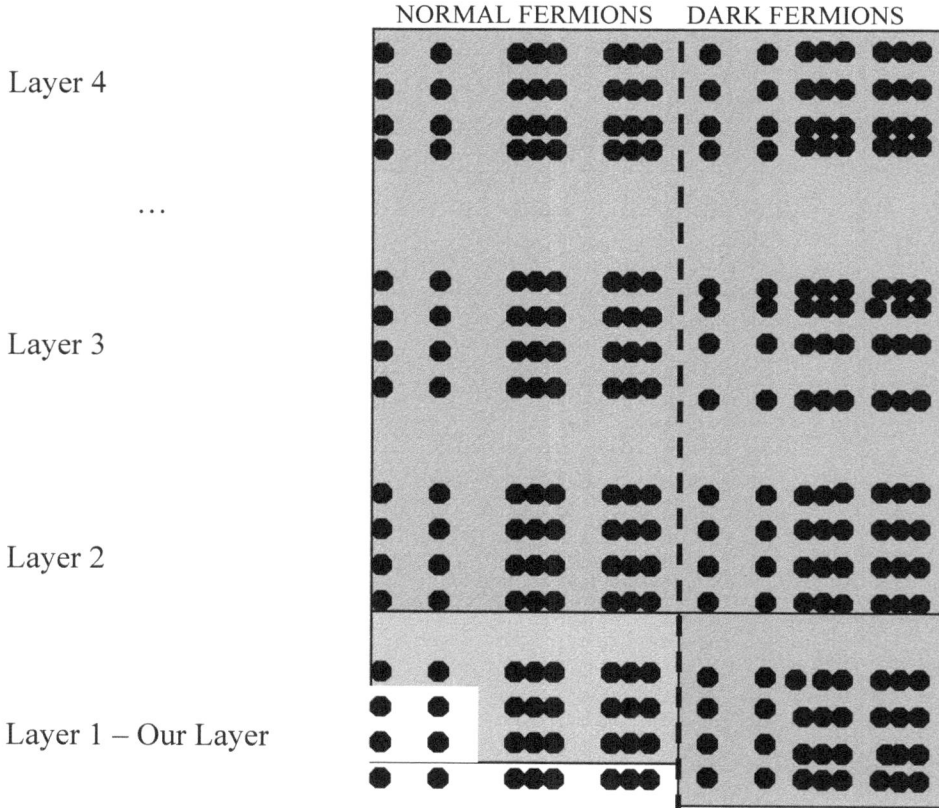

Figure 3.7 Dark and Normal fermions. Unshaded parts are the known fermions. There are 256 fundamental fermions. See Blaha (2018e).

4. Basis of Cosmos Theory

This chapter describes the basis of Cosmos Theory in totally antisymmetric tensors and then proceeds to describe the Cosmos spectrum and the dimension arrays that are generated.[12] The relations of Cosmos Theory spaces and dimension arrays with fractal curves and Dirac γ matrices are also described.

4.1 Tensor Basis

Cosmos Theory, with all its implications, is based on one mathematical fact and two assumptions. The mathematical fact follows from a consideration of the representations of the homogeneous Lorenz group in four space-time dimensions extended to other space-time dimensions r. It implies the spin of fermions is directly related to independent totally anti-symmetric tensors.

There is a difference between undeniable mathematical fact and *ad hoc* hypotheses. The fact in question is the number of *independent totally anti-symmetric tensors* of 0, 1, 2, ... , r indices in r even space-time dimensions:

$$2^{r/2} \tag{4.1}$$

This number equals the number of rows (and columns) in an irreducible γ-matrix and thus indicates its associated spin.[13] This number is twice the number of spins in even r space-time dimensions for the fermion of lowest spin.[14] The number of these spins determines the number of components of γ-matrices. Thus the number of components in an irreducible γ-matrix is

$$2^{r/2}2^{r/2} = 2^r \tag{4.2}$$

We now examine a representative set of wave functions for a fermion. We consider the form of the PseudoQuantum fermion's quantum fields:[15]

$$\psi_1(x) = \Sigma_{\alpha,s}[b_1(\alpha, s)u_{\alpha s}f_\alpha(x) + d^\dagger_1(\alpha, s)v_{\alpha s}f_\alpha^*(x)] \tag{4.3}$$
$$\psi_2(x) = \Sigma_{\alpha,s}[b_2(\alpha, s)u_{\alpha s}f_\alpha(x) + d^\dagger_2(\alpha, s)v_{\alpha s}f_\alpha^*(x)]$$

where α represents the Fourier momentum, then the total number of fermion b_i and $b_i^\dagger$ creation and annihilation operators based on counting the number of b's spins for i= 1, 2 is

---

[12] Blaha (2023d) and (2024c).

[13] For our universe where r = 4 the γ-matrices are the Dirac γ-matrices. A column number is introduced to define square γ-matrices, which are used to transform spinors.

[14] The least spin fermion quantum field in a space-time of dimension r has the spin $s = \frac{1}{2} (2^{r/2-1} - 1)$.

[15] In our PseudoQuantum formulation of fermions in the 1970's and recently we define two wave functions for each fermion for important reasons presented in my earlier papers and books.

$$2^{r/2 + 1}$$

where the additional factor of 2 arises from its PseudoQuantum formulation with two fields. If we include the anti-particle operators d_i and $d^\dagger_i$ then the total number of all operators doubles again:

$$d_{cr} = 2^{r/2 + 2} \tag{4.4}$$

where a factor of 2 is due to having two quantum fields, and a factor of 2 is from taking account of the d and $d^\dagger$ operators.

The column length d_{cr} specifies degrees of freedom since the associated creation and annihilation operators generate independent states.

4.2 Dimension Arrays

We may define an array, the associated *dimension array*, by introducing an additional index[16] creating a square array[17] with rows and columns of length d_{cr} defining a *dimension array* with d_{dr} elements:

$$d_{dr} = d_{cr}^2 = 2^{r+4} \tag{4.5}$$

This square array is analogous to a γ-matrix.

We define Cosmos spaces of dimension r, which each have an associated dimension array that specifies a set of dimensions. A dimension array specifies the spectrum of the set of fundamental fermions, of the set of scalar particles, and of the set of symmetries (including space-time dimensions r and internal symmetry dimensions.)

Each space has an even number of dimensions. Odd dimension spaces are ruled out because they would have dimension arrays that are redundant with the even space dimensions' dimension arrays. The numbers of the spins in odd dimensions are the same as the next lower even dimensions leading to redundant dimension arrays.

Dimension arrays and their associated spaces thus result.[18] We have called the theory of these spaces Cosmos Theory. Cosmos Theory is described in detail in our previous books. See Fig. 3.1 for the Physical HyperCosmos spaces.

The set of positive space-time dimension spaces ranges from r = 0 to r = ∞. We set the number of Physical spaces capable of supporting universes to ten based on the consistency condition in section 4.3 below.

We show later that the column lengths of dimension arrays form a sequence that mirrors the piecewise linear elements of the fractal Hilbert curve leading us to characterize the HyperCosmos spaces as forming a *fractal Cosmiic Curve*.

---

[16] The generation of the γ-matrices from eq. 4.1 is also done by adding a column index.

[17] A column number is introduced to define square dimension arrays that will be used later to transform dimension vectors (that are used in the study of unification.) The introduction of the column number is analogous to the introduction of a column number for γ-matrices seen on the previous page.

[18] The appearance of powers of 2 is ultimately the result of the number of independent antisymmetric tensors in each space.

4.3 Consistency Condition for Ten HyperCosmos Spaces

The set of Cosmos spaces is unlimited. However it is reasonable to consider the set of Physical HyperCosmos spaces that can support universes to be limited. Below we will consider a consistency condition that limits the number of HyperCosmos spaces to ten: from r = 0 to r = 18 in leaps of two dimensions.

The rationale for the consistency condition is described in our earlier books:[19]

The Origin of the Cosmos

We initially chose an r = 18 space as the space of the original Parent universe raises several issues. (We now prove it.)

The logic of the origin point: There can be no time before the definition of the Parent space dimension array since no space or universe existed before its definition. Yet a Parent universe cannot exist without the definition of the Parent's dimension array. Thus the dimension array and Parent universe must be simultaneously defined and created.

The only way that one may reasonably resolve this state of affairs is to define both as the consequence of a consistency condition. We suggest the consistency condition for the minimum Parent universe dimension is the requirement that its virtual thermodynamic outward pressure equals the Casimir vacuum inward pressure from its virtual external vacuum Casimir pressure at the point of origin of the virtual universe.[20] Below this minimum dimension the Parent universe cannot exist since the virtual vacuum pressure would cause it to be contracted into non-existence.

The dimension r of the Parent universe fixes the Parent universe Physical Cosmos space dimension as the least dimension whose universe is not contracting. Within the Parent universe there are child universes of lower dimension corresponding to lower dimension Cosmos spaces.

The consistency condition is based on the equality of the inner pressure of a universe and the external Casimir force. Universes where they are equal (dimension r = 18) mark the boundary between unphysical universes that collapse to a point and possible Parent universes that may expand and thus have child universes (where r < 18) are Physical. Possible Parent universes where r < 18 collapse to a point and are unphysical. *Thus the set of Physical Cosmos spaces have r ≤ 18 and are ten in number.* The consistency condition is

$$\text{Pressure} \qquad\qquad \text{Casimir Force}$$
$$N\,\Gamma((r-1)/2+1)kT/(\pi^{(r-1)/2}a^{(r-1)}) \;=\; \pi^{(r-1)/2}\,a^{-r}/[r\Gamma((r-1)/2+1)] \qquad (4.6)$$

where Cosmos Theory specifies the number of fundamental fermion species N for a universe of dimension r is $N = 2^{r+4}$.

Using Stirling's approximation and after some algebra we find eq 1.1 becomes

$$\text{Pressure Measure} \qquad\qquad \text{Casimir Force Measure}$$
$$N \qquad\qquad = \qquad\qquad (2e\pi/r)^r/(\pi a r k T) \qquad (4.7)$$

for equilibrium. If the left side of the equation were larger then universe, then expansion results. If the left side of the equation were smaller then universe, then contraction to a point results.

---

[19] Blaha (2024d, (2024e) and (2024f).
[20] Much of this chapter is based on Blaha (2022d) and (2023e).

Why Ten Physical HyperCosmos Spaces?

An examination of eq. 1.2 shows that there is a critical point in the values of N due to the factor

$$(2e\pi/r)^r \tag{4.8}$$

where the number $2e\pi/r$ is a transition point from greater than one to less than one at $r = 2e\pi = 17.08$. At this value of r and thus below the even integer dimension value $r = 18$ the Casimir force is greater than the thermodynamic pressure corresponding to a change between contraction and expansion of the Parent universe. Thus $r = 18$ marks the minimally acceptable lowest dimension Parent universe since even integer $r < 18$ would give a universe that contracts to a point and thus could not have child universes.[21] We choose the Physical Parent space dimension $r = 18$ as the minimum Physically acceptable even integer dimension. (Child universes have dimensions less than 18.)

It is also important to note that $N = 2^{r+4}$ rises with the power $r + 4$. Thus the Casimir force factor $(2e\pi/r)^r$ should also rise with the power of increasing r in the consistency condition of eqs. 4.6 and 4.7. Therefore $2e\pi/r$ must be greater than one for consistency – thus requiring the minimum Parent universe dimension r to be $[2e\pi] = 18$ where [] signifies "least integer greater than."[22]

Eq. 4.7 has been evaluated. The consistency condition[23] dimension gives $r = 18$ if

$$akT = 0.0785m_0/m = 2.98 \times 10^{-8} = 2^{-24.8} \cong 2^{-25} \tag{4.9}$$

where the radius a is set equal to $1/m$ where m is the $r = 18$ Parent universe mass, and where m_0 is the gambol mass.[24] The gambol mass is related to the universe gambol temperature T by $kT = c_g m_0$ with $c_g = 0.0785$. For $r = 18$ the consistency condition requires $m = 2.6 \times 10^6 m_0$ thus leading to eq. 4.9.

The approximate value of akT in eq. 4.9, when combined with the value of N causes the consistency condition in eq. 4.7 to become

$$\pi N2^{-25} = \pi 2^{-3} = 0.39 = (2e\pi/18)^{18} \tag{4.10}$$

---

[21] Child universes are not subject to the consistency condition and may expand as ours does.

[22] The consistency condition also suggests that powers of e and π have significance in Cosmos Theory. This point is substantiated by their appearance in coupling constant and fermion mass regularities.

[23] Blaha (2024e).

[24] See Blaha (2023e) which shows the appearance of gambol models in many particle and Cosmological cases and the importance of the gambol temperature equation.

r > 18   Expanding Universes

r = 18   ————————————Parent Universe————————————

r < 18   Universes that Contract to a Point

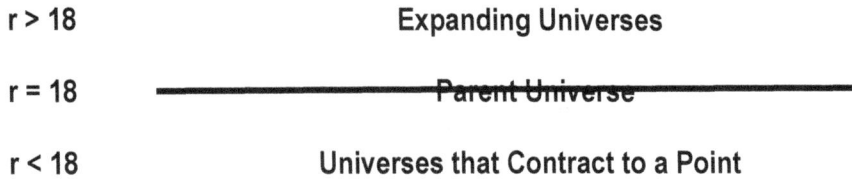

Figure 4.1 The Physical Parent Cosmos universe implied by the consistency condition. The dimension r specifies the space of the universe. It also is the space-time dimension of the universe that is allocated from within the universe space's dimension array.

The key dimension scale for universe, and thus space dimension, is 2eπ. We view this result as a confirmation of the approach in this chapter and the choice of r = 18 as the minimal possible Physical HyperCosmos Parent space for the 10 Physical HyperCosmos spaces spectrum.
The consistency condition may be related to particle interactions where an agglomeration of particles transitions to an output state. In this case the surface and volume terms suggest that interactions may be geometrically characterized and scattering amplitudes calculated geometrically. This possibility was considered in Blaha (2023e) and (2024a).

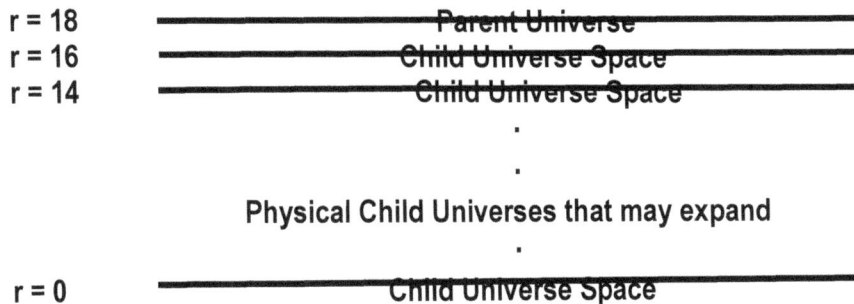

r = 18   ————————————Parent Universe————————————
r = 16   ————————————Child Universe Space————————————
r = 14   ————————————Child Universe Space————————————
                                   .
                                   .
                                   .
              Physical Child Universes that may expand
                                   .
r = 0   ————————————Child Universe Space————————————

Figure 4.2 The 10 HyperCosmos "physical" spaces that support child universes. The dimension r specifies the space of a universe. It *also* is the space-time dimension of the universe that is allocated from within the universe space's dimension array.

4.4 Extension of Cosmos Spaces to Fractional Space-time Dimensions

Cosmos spaces with fractional dimension arrays d_{dr} are part of Cosmos Theory. They are an important part of the form of the Fractal Cosmic Curve of Cosmos Theory.

The space-time dimensions that we have considered above are non-negative, even integer-valued dimensions greater than or equal to zero. We may introduce more dimensions by descending to negative space-time dimensions.[25] The spaces with

---

[25] Our ProtoCosmos model created a set of spaces with negative dimensions and fractional dimension arrays. We used them in our gambol studies (Blaha (2024a)) and to extend the Cosmos Curve to a zero dimension point. See Blaha (2024b).

negative integer dimensions are called the Limos spaces. They are used in gambol theory in Blaha (2024a).

We introduce positive (and now negative) Cayley-Dickson numbers n related to the positive (and negative) space-time dimensions r by

$$r = 2n - 2 \qquad (4.11)$$

with the result

$$d_{cn} \equiv d_{cr} = 2^{n+1} = 2^{r/2+2} \qquad (4.12)$$
$$d_{dn} \equiv d_{cr} = 2^{2n+2} = 2^{r+4} \qquad (4.13)$$

We will use n to relate the dimension arrays to a Hilbert-like fractal curve.

4.5 The Fractal Cosmos Curve

We now use the fractal relations first found in Blaha (2024b). The fractal construction of the Fractal Cosmic Curve is based on the association of dimension array column lengths[26] and the orders of piecewise linear line segment lengths that are used to form a fractal curve. We relate the *Cosmos Curve* construction to the Hilbert curve construction with:[27]

Cosmos Cayley-Dickson number $n = n_H - 1$ $\qquad (4.14)$
Hilbert Curve Line Length $= 2^{n_H}$
Hilbert Number of "boxes" $= 2^{2n_H}$
Cosmos Dimension Array Column Length $= 2^{n+1} = 2^{n_H}$
Cosmos Dimension Array Size (number of elements in array) $= 2^{2n+2} = 2^{2n_H}$

where n is the Cosmos Cayley-Dickson number and n_H is the order in the Hilbert curve construction.

The non-negative n (or r) dimension array column lengths combine to generate a two dimension filled square fractal grid from a one dimension line segment.

The negative n (r) dimension array column lengths for $n = -2$ through $n = -\infty$ combine to form a one dimension line element of length 1 equal to the array column length 1 for $n = -1$ from a zero dimension point due to the identity

$$\sum_{n=1}^{\infty} 2^{-n} = 1 \qquad (4.10)$$

If we adjoin the dimension array column lengths for $n = -\infty$ through $n = \infty$ then we have the Fractal *Cosmic Curve growing from a zero dimension point to a two dimension filled square grid.*

---

[26] Order by order (dimension by dimension) the dimension array column lengths map to the length of the corresponding Hilbert fractal curve piecewise line lengths. Blaha (2024b).
[27] The construction of the fractal curve corresponding to Cosmos Theory spaces was shown to be similar to the construction of the Hilbert fractal curve in Blaha (2024b).

The fractal dimension generation is paralleled by the generation of dimension arrays at each step in its construction.

4.6 Euclidean Construction of Creation in Cosmos Theory

With the basis of Cosmos Theory in the tensor structure of spaces we now have a complete theory with fundamental assumptions and a construction process modeled on the Fractal Cosmic Curve that parallels Euclid's formulation of Geometry. The theory is purely mathematical: a mirror of Reality.

4.7 Analogous γ-Matrix Features

The even space-time dimension γ-matrices exhibit properties similar to the Cosmos Theory dimension arrays:[28]

1. Dimension arrays have a size that is 16 times the size of the corresponding γ-matrices.

2. The γ-matrices are square. They are r matrices of size $2^{r/2}$ by $2^{r/2}$. As the even space-time dimension r increases by 2 the γ-matrices row and column sizes double. The γ-matrices quadruple as a result. For example the r = 6 γ-matrices are 8 by 8 matrices containing a quadruple of r = 4 γ-matrix parts.

3. The γ-matrices exhibit a form of nesting in quadruple multipliers.

4. Fractional γ-matrices of negative dimension n may be defined by quartering the γ-matrices of the next higher space-time dimension. These fractional matrices may be of interest for gambol quantum field theory. This topic has been considered in our previous books.

The Hilbert curve structural features, Cosmos dimension array structural features, and γ-matrix structural features are the same.

---

[28] This is discussed in detail in chapter 6.

5. Coupling Constants and Masses

5.1 Coupling Constant Values[29]

An examination of the form of eq. 4.8 suggests that the dimension $2 \cong e\pi/4$. This suggestion led the author to find a set of almost precise values of symmetry group coupling constants based on e, π and 2. Fig. 5.1 below shows an extremely regular pattern in the coupling constants of the known symmetry groups as well as U(0) – discussed later, and U(256) which appears relevant for the understanding of the gravitation constant G.

The growth in g values by factors of 2 appears to have first been noted by this author in Blaha (2018e) and (2019a) (page 58). This growth supports the quite accurate expression of values in powers of 2. Note that the values are not "bare" values but rather *renormalized values* up to small deviations comparable to experimental uncertainties.

		ESTIMATE			EXPERIMENT		
Interaction	Expression	$g^2/4\pi$ Value	g	Known[30] Value $g^2/4\pi$			Deviation
U(0) α_0	$e^2/4096$	0.0018036	-	-			-
U(1) $\alpha_1 = \alpha$	$e^2/1024$	0.00721438	0.303	0.0072973525643			1.15%
SU(2) $\alpha_2 = g^2/4\pi$	$e^2/256$	0.0289	0.63	0.0316			9.3%
SU(3) $\alpha_3 = \alpha_S$	$e^2/64$	0.115	1.21	0.117			1.7%
SU(4)[31] α_4	$e^2/16$	0.462	2.4?	0.458			0.087%

$$\alpha_n = g_n{}^2/4\pi = e^2\, 2^{2n - 12}$$

SU(256) α_{256}	$e^2 2^{500}$	2.418×10^{151}	1.74×10^{76}	-			-

Figure 5.1. Coupling Constants table from Blaha (2024h) where e = 2.718.

The coupling constants have the power representation

$$\alpha_n = g_n{}^2/4\pi = e^2\, 2^{2n - 12}$$
$$= \alpha_0\, 2^{2n} \qquad\qquad (5.1)$$

where

$$\alpha_0 = e^2/4096 = e^2 2^{-12}$$

---

[29] See Blaha (2024i) for more details.
[30] All coupling constant values are based on data from Particle Data Group Tables of 2024.
[31] This value is based on the "doubling trend" seen in the three known coupling constants g above.

5.2 Origin of the Form of Coupling Constants[32]
First we note that n is the number of fermions in a fundamental representation of SU(n). The coupling constants g have the form:

$$g(n) = e\,(4\pi)^{\frac{1}{2}}\,2^{n-6} = e\,(2^{-10}\pi)^{\frac{1}{2}}\,2^{n} = g_g 2^{n} = 0.1505\;2^{n} \tag{5.2}$$

$$= g_g\;(\text{number of spin states per fermion})^{\text{number of fermions}}$$

where e = 2.718.

The coupling constant is a fundamental value g_g times the number of spin states (a geometric property of fermions equal to one-half the column length of a spinor) raised to the number of fermions.

We see that g(n) has n factors – one factor for each 2 spin state of each fermion in the SU(n) fundamental representation. The origin of

$$g_g = e\,(2^{-10}\pi)^{\frac{1}{2}} \tag{5.3}$$

is not fully understood due to the appearance of the natural logarithm base e, which appears to have a geometric origin.

Eq. 14.3 implies that the coupling g(n) for SU(n) is g_g times n factors of 2 which represents the number of spin states per fermion, 2, in the SU(n) fundamental representation in four dimensions. Thus it is a product of degrees of freedom.

5.3 Inclusion of Gravitational Coupling Constant G
The gravitational coupling constant G of our universe can be put in the coupling constant framework that we have established.

We view the G coupling constant as embodying SU(256):

$$\alpha_{256} = g^2/4\pi = e^2 2^{500} \tag{5.4}$$

We now find the mass associated with this gravitational coupling:

$$G = g_{256}{}^2/4\pi\,M^{-2} = 0.67 \times 10^{-38}\;(\text{GeV/c}^2)^{-2} \tag{5.5}$$

We set

$$g_{256}{}^2/4\pi = \alpha_{256}$$

since the 256 fundamental fermions (both normal and Dark) in UST correspond to a SU(256) symmetry group. Then

$$M^2 = (g_{256}{}^2/4\pi)/G = 2.418{\times}10^{151}/(0.67\times10^{-38}) = 3.61 \times 10^{189}\;(\text{GeV/c}^2)^2$$

and

---

[32] Blaha (2024h).

$$M = 6 \times 10^{94} \text{ GeV/c}^2 = g^{-2} \text{ M}_{observable} = 3.5 \times 10^{13} \text{ M}_{observable} \qquad (5.6)$$

The value of $M_{observable}$ is a factor of 3.5×10^{13} times the estimated total[33] mass of the *visible* universe:[34]

$$M_{observable} = 1.71 \times 10^{81} \text{ GeV/c}^2 = 2^{269.85} \text{ GeV/c}^2$$

We now note the discrepancy between M and $M_{observable}$ has a simple representation in terms of α_0 and e. Setting

$$M_{observable} = g_M{}^2 M \qquad (5.7)$$

we find

$$g_M{}^2 \cong e^{-6}\alpha_0{}^4 = e^2(4096)^{-4} = (3.81 \times 10^{13})^{-1} = (2.62 \times 10^{-14})^{-1} \qquad (5.8)$$

with a small deviation factor of 1.089 or 8.9% in the ratio $3.81 \times 10^{13}/(3.5 \times 10^{13})$.

This use of α_0 in the study of spaces and universes is the second of five appearances in this book. It plays an important role in universe properties. We find the mass of the observable (visible) universe is proportional to the total mass of the universe based on our calculation of G.

The above result suggests that the universe is a form of particle with mass M and symmetry group SU(256). It also suggests the universe is much larger than the observable (visible) part. We have suggested this possibility several times previously in other contexts.

SU(256) like SU(3) might have been associated with confinement. However this possibility is removed by the mass factor M^{-2} that is required on dimensional analysis grounds in gravitational dynamic equations.[35] This factor causes the gravitation G to be weak.

We appear to have a viable estimate of the value of G and its origin in a coupling constant formulation.

5.4 Fermion Mass Sequences[36]

We have found a set of regularities in the fundamental fermion spectrum masses based on the appearance of powers of π and e in the consistency condition and in coupling constants. These regularities are in good approximation to the known fermion masses. Fig. 5.2 relates the masses to powers of 2, π and the base of natural logarithms e.

---

[33] This value is for normal masses - Not mass-energy..
[34] There are many suggestions of a universe mass-energy much greater than the observable.
[35] Models which change gravity from 1/r Newtonian gravity usually change the space-time dependence – not G.
[36] Blaha (2024h).

α/β	ESTIMATE	DEVIATION RATIO	DATA
1	$\nu_e = 0.0372\ ev/c^2 = 3.72 \times 10^{-11}\ GeV/c^2$		
1	$\nu' = \nu_e\, e^2 (\alpha_0^2)^{-1} = 8.4 \times 10^{-5}\ GeV/c^2 \cong e2^{-15}\ GeV/c^2$		
1	$e = r_1 \nu' = 2\,\pi\,\nu' = 0.511 \times 10^{-3}$	1.	$0.511 \times 10^{-3}\ GeV/c^2$
1	$d = r_2 e = r_2 r_1 \nu' = 16\pi\,\nu' = 4.24 \times 10^{-3}$	1.13	$3.76 \times 10^{-3}\ GeV/c^2$
¾	$u = r_3 d = r_3 r_2 r_1 \nu' = ¾ \times 2^3 \pi\,\nu' = 1.80 \times 10^{-3}$	1.02	$1.76 \times 10^{-3}\ GeV/c^2$
¾	$c = r_4 u = r_4 r_3 r_2 r_1 \nu' = ¾ \times 2^{11} \pi^2\,\nu' = 1.28$	1.001	$1.27\ GeV/c^2$
¾	$s = r_5 c = r_5 r_4 r_3 r_2 r_1 \nu' = ¾ \times 2^9 \pi\,\nu' = 102 \times 10^{-3}$	1.07	$95 \times 10^{-3}\ GeV/c^2$
½	$t = r_6 s = r_6 r_5 r_4 r_3 r_2 r_1 \nu' = ½ \times 2^{15} \pi^3\,\nu' = 171$	0.99	$172.76\ GeV/c^2$
½	$b = r_7 t = r_7 r_6 r_5 r_4 r_3 r_2 r_1 \nu' = ½ \times 2^{13} \pi\,\nu' = 4.34$	1.04	$4.18\ GeV/c^2$

Figure 5.2. Fermion masses based on α/β expressed using powers of 2, π and the base of natural logarithms e. From Blaha (2024i).

We use a sliding scale of mass dependence on the ratio of an ElectroWeak SU(2) factor and a Strong Interaction SU(3) factor. We assume they contribute with equal strength to the e and d masses; with a mild dominance of ElectroWeak over Strong for u, c, and s quarks; and with dominance of Strong over ElectroWeak by a factor of 2 for t and b quarks.

We denote the ratio of the interaction strengths with

$$\alpha/\beta \equiv \alpha_{SU(2)}(m)/\beta_{SU(3)}(m) \qquad (5.9)$$

where m is the mass of the relevant fermion. We note that the ratio of 2 for t and b reflects the ratio of the coupling constants (1.21/0.63 = 1.92 ≅ 2). The value of α/β nicely progresses from 1 to ¾ to ½ as the fermion mass increases.[37]

This suggests that Cosmos Theory is at the base of both coupling constant and fermion mass values.

5.5 Two Sequences of Fermion Masses

We have identified two sequences of fermion masses apparent in Fig. 5.2:

SEQUENCE 1				SEQUENCE 2			
e	u	c	t	ν'	d	s	b
Mass: 0.511×10^{-3}	1.80×10^{-3}	1.28	171	8.4×10^{-5}	4.24×10^{-3}	102×10^{-3}	4.34
Multiplier: $2^5 \pi$		$2^5 \pi$	$2^5 \pi$		32	32	32

where the multiplier connects adjacent masses in each sequence separately.

We suggest all four generations of fermions implement this separation into sequences.

The two sequences implement the multipliers:

---

[37] The vector and scalar boson masses are in chapter 9 and the fermion generations masses are in chapter 10 of Blaha (2024i).

Up type fermion sequence multiplier: $\mu_1 = 2^5\pi = 32\pi$ Sequence 1
Down type fermion sequence multiplier: $\mu_2 = 2^5 = 32$ Sequence 2

*We believe the multiplier of $2^5 = 32$ appearing in both sequences is **the number of fermions in a UST layer** based on the concept that all fermions of a layer are degrees of freedom for multipliers. There are 32 normal fermions in a UST layer. The factor of π for sequence 1 multiplies each of the 32 slices' mass in a fermion by a factor of π.*

Figs. 5.3 and 5.4 picture a fermion composed of an internal set of gambols (slices).

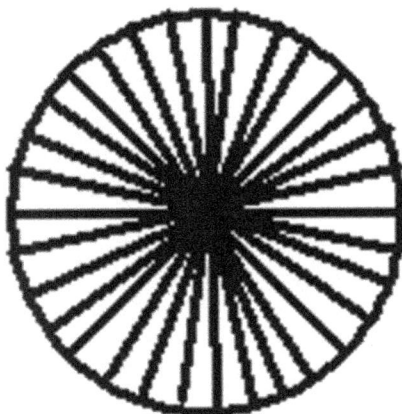

Figure 5.3. Sequence 2 fractionation of a particle into 32 gambols. Each gambol is a copy of $(2^5)^{-1}$ times the mass and structure (but not the spin or internal quantum numbers of the particle. The gambol acquires these aspects by inheritance from the particle. Each gambol acts like the particle except for gambol mass and gambol temperature.

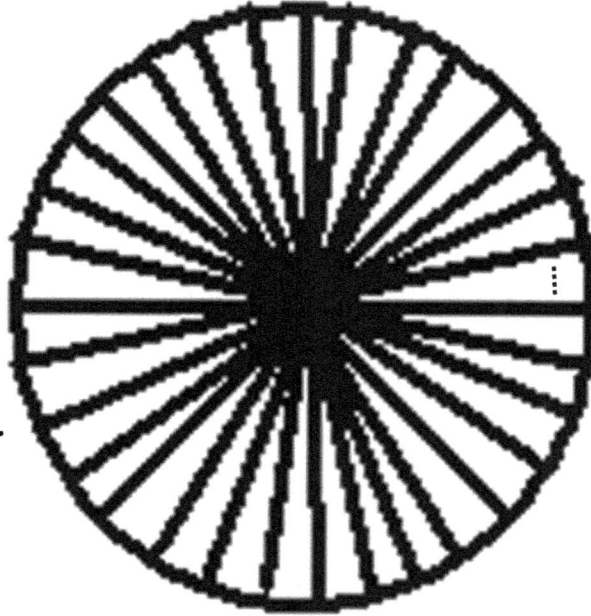

Figure 5.4. Sequence 1 fractionation of a particle into 32π gambols. Each gambol is a copy of $(\pi2^5)^{-1}$ times the mass and structure (but not the spin or internal quantum numbers) of the particle. The gambol acquires these aspects by inheritance from the particle. Each gambol acts like the particle except for gambol mass and gambol temperature.

5.6 Particle Internal Structure based on Sequences

Blaha (2024i) developed a fermion model in which fermions are mini-universes having an outer shell of a certain mass determined by a universe – particle mass spectrum, and an inner mass distribution of dimension r = 3 – perhaps composed of gambols.

The mass of the outer shell was $2^{10}3^{-5}e^{-1}\pi$ for our universe. The masses of the inner parts were denoted m_{20p} for the sequence 2 fermions where p is the particle. The masses of the inner parts were denoted m_{10p} for the sequence 1 fermions. The results are:

<u>Sequence 2 Fermion Masses</u>   (5.10)

$r = 3 \; i = 1 \quad$ gives $M_{2,3,4v'} = v' \cong e2^{-15} \; \text{GeV/c}^2 = 2^{10}3^{-5}e^{-1}\pi \; m_{20v'} = c_{\text{Shell3}} \; m_{20v'}$

$$m_{20v'} = v'/c_{\text{Shell3}}$$
$$= e2^{-15}/(2^{10}3^{-5}e^{-1}\pi) \; \text{GeV/c}^2$$
$$= (e^2/\pi) \, 2^{-25} \, 3^5 \; \text{GeV/c}^2$$

$r = 3 \; i = 2$ gives $M_{2,3,4d} = d = \pi e 2^{-11} \; \text{GeV}/c^2 = 2^{10}3^{-5} \, e^{-1}\pi \, m_{20d} = c_{\text{Shell3}} m_{20d}$

$$\begin{aligned} m_{20d} &= d/c_{\text{Shell3}} \\ &= \pi e 2^{-11}/(2^{10}3^{-5}\,e^{-1}\pi) \; \text{GeV}/c^2 \\ &= e^2 \, 2^{-21} \, 3^5 \; \text{GeV}/c^2 \end{aligned}$$

$r = 3 \; i = 3$ gives $M_{2,3,4s} = s = \tfrac{3}{4}\, \pi e 2^{-6} \; \text{GeV}/c^2 = 2^{10}3^{-5}\,e^{-1}\pi \, m_{20s} = c_{\text{Shell3}} \, m_{20s}$

$$\begin{aligned} m_{20s} &= s/c_{\text{Shell3}} \\ &= \tfrac{3}{4}\, \pi e \, 2^{-6}/(2^{10}3^{-5}\,e^{-1}\pi) \; \text{GeV}/c^2 \\ &= \tfrac{3}{4}\, e^2 \, 2^{-16} \, 3^5 \; \text{GeV}/c^2 \end{aligned}$$

$r = 3 \; i = 4$ gives $M_{2,3,4b} = b = \tfrac{1}{2}\, \pi e \; \text{GeV}/c^2 = 2^{10}3^{-5}\,e^{-1}\pi \, m_{20b} = c_{\text{Shell3}} \, m_{20b}$

$$\begin{aligned} m_{20b} &= b/c_{\text{Shell3}} \\ &= \tfrac{1}{2}\, \pi e/(2^{10}3^{-5}\,e^{-1}\pi) \; \text{GeV}/c^2 \\ &= \tfrac{1}{2}\, e^2 \, 2^{-10} \, 3^5 \; \text{GeV}/c^2 \end{aligned}$$

These four sequence 2 fermions have an inner universe shell (a sphere) of 3 dimensions. Note fermion mass is denoted by its "letter." The four values have the approximate form:

$$m_{20i} = \alpha/\beta \; e^2 2^{-31+5i} \, 3^5 \; \text{GeV}/c^2 \tag{5.11}$$
$$= \alpha/\beta \; 2^{8+5i} \, 3^5 \, e^{-1} v_e$$

for $i = 1, 2, 3, 4$.

     We now consider sequence 1 fermions, which also has internal universe shells for values of r. We set $r = 3$ again.

<u>Sequence 1 Fermion Masses</u>   (5.12)

$r = 3, \; i = 4$ gives $M_{1,3,4t} = t = \tfrac{1}{2}\, 4\pi^3 e \; \text{GeV}/c^2 = 2^{10}3^{-5}\,e^{-1}\,\pi\, m_{10t} = c_{\text{Shell3}}\, m_{10t}$

$$\begin{aligned} m_{10t} &= t/c_{\text{Shell3}} \\ &= 2\pi^3 e/(2^{10}3^{-5}\,e^{-1}\pi) = e^2\pi^2 \, 2^{-9} \, 3^5 \; \text{GeV}/c^2 \end{aligned}$$

$r = 3, \; i = 3$ gives $M_{1,3,4c} = c = \tfrac{3}{4} \times \pi^2 e 2^{-4} \; \text{GeV}/c^2 = 2^{10}3^{-5}\,e^{-1}\pi\, m_{10c} = c_{\text{Shell3}}\, m_{10c}$

$$\begin{aligned} m_{10c} &= c/c_{\text{Shell3}} \\ &= \tfrac{3}{4}\, \pi^2 e \, 2^{-4}/(2^{10}3^{-5}\,e^{-1}\pi) = \tfrac{3}{4}\, e^2\pi \, 2^{-14} 3^5 \; \text{GeV}/c^2 \end{aligned}$$

$r = 3, \; i = 2$ gives $M_{1,3,4u} = u = \tfrac{3}{4} \times \pi e 2^{-12} \; \text{GeV}/c^2 = 2^{10}3^{-5}\,e^{-1}\pi\, m_{10u} = c_{\text{Shell3}}\, m_{10u}$

$$\begin{aligned} m_{10u} &= u/c_{\text{Shell3}} \\ &= \tfrac{3}{4}\, \pi e 2^{-12}/(2^{10}3^{-5}\,e^{-1}\pi) = \tfrac{3}{4}\, e^2 2^{-22} \, 3^5 \; \text{GeV}/c^2 \end{aligned}$$

$r = 3. \; i = 1$ gives $M_{1,3,4e} = \pi e 2^{-14} \; \text{GeV}/c^2 = 2^{10}3^{-5}\,e^{-1}\pi\, m_{10e} = c_{\text{Shell3}}\, m_{10e}$

$$\begin{aligned} m_{10e} &= c/c_{\text{Shell3}} \\ &= \pi e 2^{-14}/(2^{10}3^{-5}\,e^{-1}\pi) = e^2 \, 2^{-24} 3^5 \; \text{GeV}/c^2 \end{aligned}$$

The four values have the approximate form:

$$m_{10i} = \alpha/\beta \; e^2 \, \pi^{i-2} 2^{-29+5i} \, 3^5 \; \text{GeV}/c^2 \tag{5.13}$$

$$= \alpha/\beta \; \pi^{i-2} 2^{10+5i} \; 3^5 \; e^{-1}v_e$$

for $i = 1, 2, 3, 4$.

The ratio we calculate is

$$m_{10i}/m_{20i} = 4 \; \pi^{i-2} \tag{5.14}$$
$$= M_{1,3,4i}/M_{2,3,4i}$$

Thus for b and t which both have $i = 4$:

$$M_{1,3,4t}/M_{2,3,4b} = 39.47 \tag{5.15}$$

while the experimental value is

$$t/b = 39.4$$

in good agreement.

The masses of the particle interiors m_{10i} and m_{20i} of the sequences are given by eqs. 5.11 and 5.13. In each sequence they exhibit a dependence on an integer labeled i. We now treat those sequence numbers that specify each particle in each sequence and denote it as n:

$$m_{20n} = \alpha/\beta \; e^2 2^{-31+5n} \; 3^5 \; GeV/c^2 \tag{5.16}$$
$$= \alpha/\beta \; 2^{8+5n} \; 3^5 \; e^{-1}v_e$$
$$= \alpha/\beta \; (2^5)^n \; 2^8 \; 3^5 \; e^{-1}v_e$$

$$m_{10n} = \alpha/\beta \; e^2 \; \pi^{n-2} 2^{-29+5n} \; 3^5 \; GeV/c^2 \tag{5.17}$$
$$= \alpha/\beta \; \pi^{n-2} 2^{10+5n} \; 3^5 \; e^{-1}v_e$$
$$= \alpha/\beta \; (2^5\pi)^n \; 2^{10} \; 3^5 \pi^{-2} e^{-1}v_e$$

for $n = 1, 2, 3, 4$. We further introduce a "particle" entry into each sequence for the case of $n = 0$:

$$m_{200} = e^2 2^{-31} \; 3^5 \; GeV/c^2 \tag{5.18}$$
$$= 2^8 \; 3^5 \; e^{-1}v_e = 2.63 \times 10^{-6} \; GeV/c^2$$

$$m_{100} = e^2 \; \pi^{-2} 2^{-29} \; 3^5 \; GeV/c^2 \tag{5.19}$$
$$= 2^{10} \; 3^5 \; \pi^{-2} e^{-1}v_e = 5.08 \times 10^{-6} \; GeV/c^2$$

with $\alpha/\beta = 1$ by extending the particle list in Fig. 5.2 to the $n = 0$ case for each sequence.

5.7 Derivation of the Form of Internal Fermion Masses

Earlier we derived the form of the Coupling Constants. We found their form followed from the number of fermions in a fundamental representation of SU(n). A coupling constant $g(n) = g_n$ has the form of eq. 5.2.

We can now apply a similar logic to define the forms of m_{10n} and m_{20n}. In the case of sequence 2 we ascribe the factor of $(2^5)^n$ to the number of Normal fermions in one UST layer raised to the power of n:

$$(2^5)^n = \text{(number of Normal fermions in one layer)}^n = 32^n \qquad (5.20)$$

where n indicates the place of the fermion in sequence 2. *This factor reflects the structuring of Cosmos Theory applied to the UST as a 32 dimension layer mapped to fermions.*

In the case of sequence 1 we approximate the factor of $(2^5\pi)^n \cong (2^7)^n$ where 2^7 is the total number of Normal fermions for all four UST layers. *This factor reflects the structuring of Cosmos Theory applied to the UST as four layers of $2^7 = 128$ dimensions mapped to fermions.*

We raise this number to the power n:

$$(2^5\pi)^n \cong (2^7)^n = \text{(number of Normal fermions in all four layer)}^n = 128^n \qquad (5.21)$$

where n indicates the place of the fermion in sequence 1.

Thus we find that fermion masses build in multiples, $(2^5)^n$ or $(2^5\pi)^n$, of the previous fermion mass, fermion by fermion, in each sequence in a manner analogous to the growth of coupling constants, factor by factor of fermion spin value $2 = 2\times\frac{1}{2} + 1$, for each coupling constant.

Note: The product of the sequence multipliers appears related to the coupling constant value g_g:

$$g_g = e\,(2^{-10}\pi)^{\frac{1}{2}} \qquad (5.22)$$
$$= e\pi/(\mu_1\mu_2)^{\frac{1}{2}}$$

where $\mu_2 = 2^5$ and $\mu_1 = 2^5\pi$.

The Coupling Constants, the dimension arrays, and the relation of coupling constants to masses (and thereby to energies) all reflect their common interpretation as manifestations of degrees of freedom within Cosmos Theory. There is a unifying principle at the basis of Cosmos Theory in its spaces, symmetries, universes, universe structure, energies, and particle interactions.

6. Dynamical Analysis of Dimension Arrays

This chapter shows an intimate connection between Cosmos spaces dimension arrays, the Quadplex structure of wave functions including tachyonic aspects, and the matrices (γ-matrices, SU(4) matrices, and SU(2)⊗U(1) matrices), which appear in the dynamic equations for fermions in our universe. Cosmos Theory may be said to "explain it all" for elementary particle structure and dynamics.

Cosmos dimension arrays are the template for fundamental particle matrix dynamics.

The dynamical analysis of dimension arrays leads to a new and exciting connection between Cosmos Theory and the Quadplex unification formalism for the ElectroWeak and Strong interactions in the superluminal and subluminal quantum field theory presented in Blaha (2024j).

Below we show the need to use Quadplex wave functions for fundamental particles to facilitate the use of the dimension arrays of Cosmos Theory. We also need a PseudoQuantum wave function formalism in Two-Tier Theory to achieve the unification of all symmetries in all possible Cosmos universes. We defer using a PseudoQuantum formalism for the present in the interests of having a clear presentation since the Quantum formulation presented here that brings out the ideas embodied in Cosmos dimension arrays.

In addition, in the interests of clarity, we also do not use the author's Two-Tier Theory[38] in this chapter. It is required to achieve finite perturbation theory results in four dimensions (our universe) and universes of higher space-time dimension.[39] We view Two-Tier Theory[40] as the only viable way to have finite perturbation theory results to all orders in all possible universes of any dimension. Its use in Cosmos Theory makes it the only "universe friendly" *fundamental* Theory of Everything. The ramifications of the theory such as the Hubble expansion of universes remain to be fully understood.

The set of Dirac- γ-matrices in our universe appear as 16 submatrices in a 256 dimension array in a Quadplex formulation of fermion dynamic equations.

The 16 matrices in the SU(4) fundamental <u>4</u> representation appear as 16 submatrices in a 256 dimension array.

The 4 matrices of SU(2)⊗U(1) appear within four fundamental representations along the diagonal of a 256 dimension array.

---

[38] See Blaha (2002).
[39] Other approaches to renormalization do not work in general in higher dimension quantum field theories.
[40] Two-Tier Theory "adds" two imaginary q-number dimensions to four dimension space-time curing perturbation theory of infinities. Two-Tier Theory introduces an imaginary q-number part to all coordinates y, z, u, v. See Blaha (2024j). While the use of Two-Tier Theory gives the conventional perturbation theory results at low energies, it gives new behavior at ultra-high energies.

The required Quadplex form of fermion wave functions necessarily brings in bradyon-tachyon wave functions in a manner consistent with the interactions. *Thus an aspect of faster than light motion – tachyons – is natural to the dynamics of Cosmos Theory matter.*

6.1 The Cosmos Dimension Arrays

Each of the ten physical Cosmos spaces (Fig. 6.1) has an associated dimension array. In universes of each Cosmos space type the dimension array maps to the spectrum of fundamental fermions and to the set of internal symmetry groups. See Figs. 3.6 and 3.7.

In this chapter we show that the dimension array of a space specifies dynamical Dirac equation features as well:

1. The form of the derivative term γ matrices for fermion dynamic equations (Dirac equations).

2. The forms of the SU(4) interaction terms representations. (Also SU(3)⊗U(1) after breakdown).

3. The forms of the SU(2)⊗U(1) interaction terms representations.

An important aspect of these features is that they all require the Quadplex formalism developed by this author in Blaha (2024j). This formalism combines bradyon (subluminal) and tachyon (superluminal) parts within each fundamental fermion. It opens the door to faster than light Physics. It also has an intriguing side effect: It can account for the instantaneous nature of Quantum Entanglement through the tachyon part of Quadplex wave functions.

The following sections show an embedding of Quadplex wave function features in dimension arrays in the Unified SuperStandard Theory (UST) developed by this author in recent years.

The remaining sections for the UST may be directly extended to the case of dimension arrays for higher dimension (spaces) universes such as the 6 dimension Megaverse.

6.2 Dirac-like Field Equation

For a *free* Quadplex wave function in our universe we have

$$(i \, {}^{y}\gamma^{\mu} \, \partial/\partial y^{\mu} + {}^{z}\gamma^{\mu} \, \partial/\partial z^{\mu} + i \, {}^{u}\gamma^{\mu} \, \partial/\partial u^{\mu} + {}^{v}\gamma^{\mu} \, \partial/\partial v^{\mu} + M)\psi(y, z, u, v) = 0 \quad (6.1)$$

where μ is the space-time index. The equation may be reduced to subsidiary equations that are combinations of bradyon B and tachyon T subsidiary equation terms as in Blaha (2024j). The combinations are composed of parts of the Quadplex wave functions

$(B_yB_z, B_yT_z, T_yB_z, T_yT_z)(B_uB_v, B_uT_v, T_uB_v, T_uT_v)$ where the susbscripts indicate the y, z, u, v coordinate syatems. Blaha (2024j) points out:

> "We will again use the distinction between bradyons and tachyons to create a set of four Quadplex fields. Each Quadplex fermion field will consist of four parts where each part is either bradyonic or tachyonic. We associate each Quadplex field with a fermion.
> We signify a bradyonic field with b, and a tachyonic field with t. We will use the up-type and down-type[41] fermion sequences found in our previous books for Quadplex fields:

Up-Type Sequence 1

$$e \leftrightarrow \psi_e = \psi_{1\uparrow} \sim b_y b_z t_u t_v \qquad (6.2)$$
$$u \leftrightarrow \psi_u = \psi_{2\uparrow} \sim b_y t_z t_u b_v$$
$$c \leftrightarrow \psi_c = \psi_{3\uparrow} \sim b_y b_z b_u t_v$$
$$t \leftrightarrow \psi_t = \psi_{4\uparrow} \sim b_y b_z b_u b_v$$

Down-Type Sequence 2

$$v' \leftrightarrow \psi_{v'} = \psi_{1\downarrow} \sim b_y t_z t_u t_v$$
$$d \leftrightarrow \psi_d = \psi_{2\downarrow} \sim b_y t_z t_u b_v$$
$$s \leftrightarrow \psi_s = \psi_{3\downarrow} \sim b_y t_z b_u t_v$$
$$b \leftrightarrow \psi_b = \psi_{4\downarrow} \sim b_y t_z b_u b_v$$

where we use particle symbols to represent wave functions with coordinates indicated with subscripts. The ElectroWeak part of each fermion uses y and z coordinates, and the SU(4) part uses u and v coordinates (together with y and z parts). All four coordinate systems have four space-time dimensions. The top-type fermion assignment [in eq. 6.2] is *generally based on the ordering of quark masses, with heavier fermions having more Bradyon parts and lighter fermions having more Tachyon parts.* A similar assignment may be made for down-type fermions. The list is modified to match ElectroWeak pairs with up-type bradyons corresponding down-type tachyons except for the e – v (neutrino) case where the match is e - v'."

We will use the Quadplex formalism below. *The Quadplex formalism is naturally appropriate for the Cosmos Theory – UST formulation.*

6.3 Derivative Terms

Free fermion wave functions $\psi(y, z, u, v)$ have a general derivative form:

$$i\; {}^y\gamma^\mu \, \partial/\partial y^\mu + {}^z\gamma^\mu \, \partial/\partial z^\mu + i\; {}^u\gamma^\mu \, \partial/\partial u^\mu + {}^v\gamma^\mu \, \partial/\partial v^\mu \qquad (6.3)$$

described in Blaha (2024j). The omission of i's follows from tachyonic-like part behavior. Other forms with a different pattern of i's are also used in Blaha (2024j) as indicated in the above quote.

The γ matrices for the coordinate systems are shown to be mapped to a UST dimension array in Fig. 6.1. Note that the Quadplex formalism is needed to "fill" the dimension array with γ matrices.

---

[41] Up-type particles are isospin up particles. Down-type particles are isospin down particles.

The γ matrices embedding in a dimension array for the UST may be directly extended to the case of dimension arrays for higher dimension (spaces) universes such as the 6 dimension Megaverse using the quadrupling mechanism for dimension arrays.

6.4 SU(4) Interaction Term

The Strong Interaction SU(4) group has a four complex coordinates fundamental representation. We will define two representations. One representation (sequence 1) will describe up-type fermions: v', u, c, and t. The other representation (sequence 2) will describe down-type fermions: e, d, s, and b. These sequences were described in Blaha (2024i). The v' fermion is a heavy neutrino corresponding to the first generation v. It has not yet been found in Nature. The separation of the set of fermions into two sequences is based in part on their differing electric charges with down-type fermions one charge unit below their corresponding up-type fermion.

We define the SU(4) term in the fermion field equation with

$$g_s\ A^{a\mu}(y)\ ^y\gamma_\mu T_a \tag{6.4}$$

where the T_a are 4×4 SU(4) matrices, $^y\gamma_\mu$ are Dirac matrices for the y coordinate space, g_s is the SU(4) coupling constant, and $A^{a\mu}(y)$ represents SU(4) vector boson fields. (This definition can be generalized to embody other coordinate systems. For example, $A^{a\mu}(y, z, u, v)$ is a theoretic possibility although this type of extension is not supported by experiment.)

The SU(4) interaction term becomes different if we use the bradyon-tachyon formulation of SU(4) interactions within the framework of a bradyon-tachyon formalism developed in chapter 9 of Blaha (2024i). In this theory fermions are endowed with bradyon and tachyon parts. Eq. 6.2 shows the y, z, u and v coordinate system parts of the eight first generation fermions. Wave functions have transitions between bradyon and tachyon parts under SU(4) and ElectroWeak transformations. The SU(4) matrices T_{kij} must be generalized to have factors within them producing bradyon-tachyon changes. The SU(4) matrices are replaced by $[S_{kij}]$ matrices:

$$S_{kij} = T_{kij} T^S_{ij} \tag{6.5}$$

Under SU(4) u and v based bradyon-tachyon changes we found the T^S_{ij} form

$$T^S = T^S_\uparrow = T^S_\downarrow = \begin{bmatrix} 1 & S_v & S_u & S_u S_v \\ S_v^{-1} & 1 & S_u S_v^{-1} & S_u \\ S_u^{-1} & S_u^{-1} S_v & 1 & S_v \\ S_u^{-1} S_v^{-1} & S_u^{-1} & S_v^{-1} & 1 \end{bmatrix} \tag{6.6}$$

in Blaha (2024i) based on its section 6.4.3 where

$$S_u \psi(u) = \psi_T(u)$$
$$S_u^{-1} \psi_T(u) = \psi(u)$$
$$S_v \psi(v) = \psi_T(v)$$
$$S_v^{-1} \psi_T(v) = \psi(v)$$

maps between bradyons and tachyons. The minimal transformations are[42]

$$S_u = i\gamma_u{}^0 \gamma_{u3}$$
$$S_v = i\gamma_v{}^0 \gamma_{v3}$$
$$S_u^{-1} = -i\gamma_u{}^0 \gamma_{u3}$$
$$S_v^{-1} = -i\gamma_v{}^0 \gamma_{v3}$$

    Fig. 6.2 shows the SU(4) T_k matrices embedded within a UST dimension array. *Note that the Quadplex formalism is needed to map the dimension array to SU(4) generator matrices.*

    The T matrices mapped to a dimension array for the UST may be directly extended to the case of dimension arrays for higher dimension (spaces) universes such as the 6 dimension Megaverse. This mapping is facilitated by the mapping of the dimension r = 4 dimension array in quadruplicate in the r = 6 dimension array and so on.

6.5 Dimension Array Quadplex SU(4) S Matrices

    The Quadplex form of the SU(4) interaction using eq. 6.5 is

$$g_s \, A^{a\mu}(y) \, {}^y\gamma_\mu S_a \tag{6.7}$$

This form would also describe the SU(3)⊗U(1) interaction after breakdown.

6.6 ElectroWeak SU(2)⊗U(1) Interaction Terms

    The ElectroWeak Interaction SU(2)⊗U(1) group has a two fermion fundamental representation. Each representation contains a corresponding pair of up-type and down-type fermions. They were described in Blaha (2024i).

    Each of the ElectroWeak representations has a pair of fields:

$$\psi_{EW}(y, z) = \begin{bmatrix} \psi_{T_z}(y, z) \\ \\ \psi(y, z) \end{bmatrix}$$

with the top component having a tachyonic z wave function *part* (consistent with the ordering of Figs. 6.6 and 6.7, and the bottom component having a bradyonic z wave

---

[42] Section 9.2 of Blaha (2024i).

function *part*. The combined one generation, one layer wave function in Fig. 6.7 can be represented by the set of Quadplex wave functions:

$$\Psi_{EW} = (\psi_{EW1}(y, z), \psi_{EW2}(y, z), \psi_{EW3}(y, z), \psi_{EW4}(y, z))$$

We define the form of ElectroWeak interaction terms in the fermion field equation with

$$ig_{EW}\mathbf{t}(S_z)W^{\mu}(y) + ig_{EW}'t_0(S_z)W_0{}^{\mu}(y) \tag{6.8}$$

using S_{EW} matrices where $W^{a\mu}(y)$ represents the $SU(2) \otimes U(1)$ vector boson fields, and g_{EW} and g_{EW}' are the coupling constants.

   The Pauli matrices that appear in the ElectroWeak Lagrangian must be enhanced to incorporate bradyon-tachyon transitions due to the different forms of the fermion fields: bradyon-bradyon charged leptons and bradyon-tachyon neutral leptons.

   The enhanced $\mathbf{t}(S_z)$ and $t_0(S_z)$ Pauli matrices depend on transformations between the bradyon and tachyon z part of wave function pairs. They have the form[43]

$$t_- = \tfrac{1}{2} \begin{bmatrix} 0 & S_z\gamma^5 \\ 0 & 0 \end{bmatrix} \tag{6.39}$$

$$= \tfrac{1}{2} \begin{bmatrix} 0 & i\gamma_z{}^0\gamma_{z3}\gamma_z{}^5 \\ 0 & 0 \end{bmatrix}$$

$$t_+ = \tfrac{1}{2} \begin{bmatrix} 0 & 0 \\ S_z & 0 \end{bmatrix} \tag{6.40}$$

$$= \tfrac{1}{2} \begin{bmatrix} 0 & 0 \\ i\gamma_z{}^0\gamma_{z3} & 0 \end{bmatrix}$$

---

[43] Using the equation numbers and notation of Blaha (2024i).

$$t_3 = \tfrac{1}{2} \begin{bmatrix} -i\gamma_z{}^0\gamma_{z3}\gamma_z{}^5 & 0 \\ 0 & i\gamma_z{}^0\gamma_{z3} \end{bmatrix} \tag{6.41}$$

$$\equiv \tfrac{1}{2} \begin{bmatrix} -1 & 0 \\ 0 & 1 \end{bmatrix}$$

$$t_0 = \tfrac{1}{2} \begin{bmatrix} 1 & 0 \\ 0 & 1 \end{bmatrix} \tag{6.42}$$

using

$$S_z = i\gamma_z{}^0\gamma_{z3}$$
$$S_z{}^{-1} = -i\gamma_z{}^0\gamma_{z3}$$

described in Blaha (2024i). The electric charge is

$$Q = t_0 + t_3 \tag{6.43}$$

and the interaction terms for each of the four representations in Fig. 6.7 become

$$t(S_z)\cdot W^i(y) \rightarrow t_- W^+ + t_+ W^- + t_3 W^3 \tag{6.8'}$$
$$t_0(S_z) W_0{}^0(y) \rightarrow t_0 W^0$$

Fig. 6.7 shows the four SU(2)⊗U(1) representations embedded, in diagonal form, within a UST dimension array. There are four pairs of fermions within the eight fermions in each generation in each layer of the UST fermion spectrum.

Each column, of the four columns, in Fig. 6.3 corresponds to a representation of SU(2)⊗U(1) denoted with an upper index. The four entries in each column are its SU(2)⊗U(1) 4 × 4 submatrices denoted with lower indices for the real-valued SU(2)⊗U(1) representations.

The three other UST layers have different SU(2)⊗U(1) groups and representations. The submatrices of these groups have the same form as the Fig. 6.9 submatrices.

The matrices map for a dimension array for the UST may be directly extended to the case of dimension arrays for higher dimension (spaces) universes such as the 6 dimension Megaverse. This map is facilitated by the embedding of the dimension $r = 4$ dimension array in quadruplicate in the $r = 6$ dimension array, and so on in higher dimension arrays.

Note that the Quadplex formalism is needed to "fill" the dimension array with $t(S_z)$ and $t_0(S_z)$ *matrices in Fig. 6.3.* These matrices are used in the ElectroWeak ordering of Figs. 6.6 and 6.7.

6.7 Mass Terms

The sequences listed in eq. 6.2 were first found in an examination of physical fundamental fermion masses in the first generation fermions in the UST. The sequences are described in detail in Blaha (2024i).

The sequences and their masses are

	Sequence 1				Sequence 2		
	e	u	c	t	ν'	d	s	b
Mass (GeV/c²):	0.511×10^{-3}	1.80×10^{-3}	1.28	171	8.4×10^{-5}	4.24×10^{-3}	102×10^{-3}	4.34
Multiplier:		$2^5 \pi$	$2^5 \pi$	$2^5 \pi$		$2^5 = 32$	$2^5 = 32$	$2^5 = 32$

Each sequence has a multiplicative relation between the masses: 2^5 for sequence 2 and $2^5 \pi$ for sequence 1.

The diagonal Lagrangian fermion mass term appears in Fig. 6.8.

6.8 General Form of the Strong Interaction Lagrangian

The form of the 8 dimension matrix composed of the pair of 4 dimension fermion reprsentations $\mathcal{L}_{Strong}$ of the Strong Interaction SU(4) Lagrangian appears in Fig. 6.5. (The ElectroWeak Lagrangian is discussed separately in section 6.9 to avoid a confusing use of wave function and interaction indices.) The solely Strong Lagrangian is:

$$\overline{\Psi}[i\,^y\gamma^\mu\,\partial/\partial y^\mu + {}^z\gamma^\mu\,\partial/\partial z^\mu + i\,^u\gamma^\mu\,\partial/\partial u^\mu + {}^v\gamma^\mu\,\partial/\partial v^\mu]\Psi + \mathcal{L}_{Mass} + \mathcal{L}_{Strong} \qquad (6.9)$$

with Ψ defined in Fig. 6.4, $\mathcal{L}_{Mass}$ defined in Fig. 6.8, and $\mathcal{L}_{Strong}$ defined in eq. 6.7 and Fig. 6.8.

6.9 Form of the ElectroWeak Interaction Lagrangian Terms

The solely ElectroWeak Lagrangian can be put in an eight dimension matrix form. We define an 8-vector representing the up-type and down-type sequences of fermions ordered to be in ElectroWeak pairs as in Fig. 6.6.

$$\overline{\Psi}_{EW}[i\,^y\gamma^\mu\,\partial/\partial y^\mu + {}^z\gamma^\mu\,\partial/\partial z^\mu + i\,^u\gamma^\mu\,\partial/\partial u^\mu + {}^v\gamma^\mu\,\partial/\partial v^\mu]\Psi_{EW} + \mathcal{L}_{MassEW} + \mathcal{L}_{ElectroWeak} \qquad (6.10)$$

with Ψ_{EW} defined in Fig. 6.6, $\mathcal{L}_{MassEW}$ defined in Fig. 6.9, and $\mathcal{L}_{ElectroWeak}$ defined in Fig. 6.7.

The ElectroWeak Lagrangian is symbolized by a diagonal matrix of 2×2 submatrices in Fig. 6.7. These submatrices correspond, submatrix by submatrix, with the order of the fermion wave function vector of Fig. 6.6.

6.10 Cosmos Theory Dimension Arrays Dovetail with Quadplex Wave Functions

The preceding discussion, and the figures that follow, show that the UST dimension array for our universe is compatible with ElectroWeak $SU(2) \otimes U(1)$ and Strong Interaction $SU(4)$ (and the broken $SU(3) \otimes U(1)$) formalisms – Figs. 6.2 – 6.9. One sees the $SU(4)$ fundamental representation matrices fit directly in the 256 dimension UST dimension array. One also sees the set of $SU(2) \otimes U(1)$ fundamental representation matrices fit directly in the 256 dimension UST dimension array.

Thus we may regard the Quadplex (and duplex) wave function formalisms, which support tachyon physics, as bringing "hidden" tachyonic behavior to Cosmos Theory. The Lagrangian terms for the Strong and ElectroWeak Interactions may be viewed as directly based on Cosmos Theory dimension array diagrams.

Cosmos dimension arrays are the template for fundamental particle dynamics.

The Quadplex and duplex formalisms for the UST in 4 dimensions generalize directly to higher Cosmos dimensions due to the quadrupling mechanism within HyperCosmos as dimension increases by twos. Thus a form of "hidden" tachyonic behavior exists in all universes of all Cosmos Theory spaces.

${}^{y}\gamma^{0}$	${}^{z}\gamma^{0}$	${}^{u}\gamma^{0}$	${}^{v}\gamma^{\mu 0}$
${}^{y}\gamma^{1}$	${}^{z}\gamma^{1}$	${}^{u}\gamma^{1}$	${}^{v}\gamma^{1}$
${}^{y}\gamma^{2}$	${}^{z}\gamma^{2}$	${}^{u}\gamma^{2}$	${}^{v}\gamma^{2}$
${}^{y}\gamma^{3}$	${}^{z}\gamma^{3}$	${}^{u}\gamma^{3}$	${}^{v}\gamma^{3}$

Figure 6.1. The dimension array for the γ matrices of the derivative terms. There is one subarray for each space-time index of the r = 4 UST of our universe. The y, z, u, and v indices indicate the four Quadplex coordinate systems. Each of the 16 γ submatrices is a 4 × 4 Dirac γ matrix. The ${}^{y}\gamma^{\mu}$ matrix is for the coordinate system of our universe with μ = 0, 1, 2, and 3.

T^1	T^2	T^3	T^4
T^5	T^6	T^7	T^8
T^9	T^{10}	T^{11}	T^{12}
T^{13}	T^{14}	T^{15}	I

Figure 6.2. The dimension array for the 16 U(4) T_k matrices of the SU(4) interaction terms of the r = 4 UST of our universe. Each of 15 submatrices is a 4 × 4 SU(4) matrix. The 16th submatrix is the identity matrix. These matrices are for the first UST layer. Other UST layers have different SU(4) groups. The submatrices of these groups have the same form as the above submatrices. The dimension array supports a map to the set of U(4) matrices.

$T_{EW1}{}^1$	$T_{EW1}{}^2$	$T_{EW1}{}^3$	$T_{EW1}{}^4$
$T_{EW2}{}^1$	$T_{EW2}{}^2$	$T_{EW2}{}^3$	$T_{EW2}{}^4$
$T_{EW3}{}^1$	$T_{EW3}{}^2$	$T_{EW3}{}^3$	$T_{EW3}{}^4$
$T_{EW4}{}^1$	$T_{EW4}{}^2$	$T_{EW4}{}^3$	$T_{EW4}{}^4$

Figure 6.3. There are four pairs of fermions for the eight fermions in the four UST generations of the first layer as one can see from the sequences in eq. 6.2. Each column corresponds to a representation of SU(2)⊗U(1) denoted with an upper index. The four entries in each column are its SU(2)⊗U(1) 4 × 4 submatrices denoted with lower indices in its real-valued SU(2)⊗U(1) representation. The other UST layers have different SU(2)⊗U(1) groups and representations. The submatrices of these groups have the same form as the above submatrices. Dimension arrays support maps to the set of U(4) matrices in any Cosmos space.

$$\Psi = \begin{bmatrix} e \\ u \\ c \\ t \\ v' \\ d \\ s \\ b \end{bmatrix} \begin{array}{l} \text{Sequence 1} \\ \\ \\ \text{Sequence 2} \end{array}$$

Figure 6.4. The wave functions of the eight first generation, first layer UST fundamental fermions for use in the Strong Lagrangian ordered by their sequences and positions within the sequences. The other wave functions of the other generations and layers are similar in form. The wave functions are denoted by the fermion's acronym.

$$\mathcal{L}_{\text{Strong}} = \overline{\Psi} \begin{bmatrix} \begin{array}{c} \text{SU(4)} \\ \text{Sequence 1} \end{array} & 0 \\ 0 & \begin{array}{c} \text{SU(4)} \\ \text{Sequence 2} \end{array} \end{bmatrix} \Psi$$

Figure 6.5. Eight dimension representation of the pair of four dimension SU(4) representations of the fermion sequences in generation 1, layer 1 of Normal fermions.

$$\Psi_{EW} = \begin{bmatrix} v' \\ e \\ d \\ u \\ s \\ c \\ b \\ t \end{bmatrix}$$

Figure 6.6. Fermion wave functions terms ordered as four sets of fermion wave function pairs for ElectroWeak use use in Fig. 6.7.

$$\begin{bmatrix} SU(2) \otimes U(1) & & & \\ & SU(2) \otimes U(1) & & \\ & & SU(2) \otimes U(1) & \\ & & & SU(2) \otimes U(1) \end{bmatrix}$$

Figure 6.7. Four ElectroWeak representations for one generation of one layer. The representations correspond with the order of the fermions in the fermion wave function vector of Fig. 6.6.

$$\mathcal{L}_{Mass} = \overline{\Psi} \begin{bmatrix} e & & & & \text{Sequence 1} & & \\ & u & & & & & \\ & & c & & & & \\ & & & t & & & \\ & & & & v' & & \\ & & & & & d & \\ & \text{Sequence 2} & & & & & s & \\ & & & & & & & b \end{bmatrix} \Psi$$

Figure 6.8. Diagonal Lagrangian fermion mass matrix for generation 1, layer 1 of the Strong SU(4). Masses are denoted by fermion symbol.

$$\mathscr{L}_{\text{MassEW}} = \overline{\Psi}_{\text{EW}} \begin{bmatrix} \nu' & & & & & \\ & e & & & & \\ & & d & & & \\ & & & u & & \\ & & & & s & \\ & & & & & c & \\ & & & & & & b \\ & & & & & & & t \end{bmatrix} \Psi_{\text{EW}}$$

Figure 6.9. Diagonal Lagrangian fermion mass matrix for generation 1, layer 1 of the ElectroWeak Lagrangian. Masses denoted by fermion symbol. Other components are zeroes (not shown). The dimension array supports a map to this set of ElectroWeak mass matrices.

7. Generation, Layer and Connection Groups Matrices

The Generation and Layer groups were introduced by the author some years ago as an extension of The Standard Model, which we called the Unified SuperStandard Theory (UST). Chapter 6 presented the Standard Model internal symmetry groups for the ElectroWeak and Strong interactions. This chapter presents the Generation, Layer and Connection groups. The introduction of these groups implies new interaction terms in the fermion Dirac equation. In addition to the Strong interaction term in eq. 6.7 there are Generation and Layer group Lagrangian interaction terms:

$$g_G \, A_G{}^{a\mu}(y) \, {}^y\gamma_\mu S_{Ga} \; + g_L \, A_L{}^{a\mu}(y) \, {}^y\gamma_\mu S_{La} \qquad (7.1)$$

as well as Connection group terms in a Quadplex formulation similar to that of chapter 6. The subscript "G" labels the Generation group interaction. The subscript "L" labels the Layer group interaction. Fig. 7.2 displays the Generation and Layer group U(4) generator matrices mapped from the UST dimension array just as Strong SU(4) matrices are mapped from the UST dimension array in Fig. 6.2. The S_{Ga} and S_{La} matrices have forms similar to those of chapter 6:

$$S_{Gkij} = T_{Gkij} T_G{}^S{}_{ij} \qquad (7.2)$$
$$S_{Lkij} = T_{Lkij} T_L{}^S{}_{ij}$$

where, under SU(4) u and v based bradyon-tachyon changes, we found the $T_G{}^S{}_{ij}$ and $T_L{}^S{}_{ij}$ have forms similar to eq. 6.6.

The terms in eq. 7.1 have T_{Gkij} and T_{Lkij} matrices that span the entire set of fermion layers. We will discuss them in detail in chapter 8.

These additional interactions also imply the Cabibbo-Kobayashi-Maskawa (CKM) matrix must be generalized to accommodate these interactions for all UST layers.

Recently experimental evidence has surfaced suggesting the CKM matrix must be modified to accommodate new interactions beyond those of the Standard Model. We suggest these new interactions may be our Generation and Layer group interactions.

The Generation group representations would directly cause modifications in the CKM matrix for the first UST layer.

The Layer group representations also would directly cause modifications in the CKM matrix for the first UST layer.

Since the Layer group fermion representations mix fermions of all layers the CKM matrix would have to be generalized to apply to all UST fermion layers.[44]

7.1 The Number of Groups in the UST of our Universe

The 256 fundamental fermions mapped from the Cosmos UST dimension array of our universe occupy four layers. The lowest mass fermions generation appears in layer 1. Each species[45] has a representative fermion in this generation. See Fig. 7.3.

Fig. 7.1 shows 8 Generation groups in the UST. Four groups are in the Normal sector. We placed four groups in the Dark sector based on an assumed similarity between Normal and Dark sectors implicit in the UST dimension array. The 8 Generation groups are independent of each other.

Fig. 7.1 also shows 8 Layer groups in the UST. Four groups are in the Normal sector. We place four groups in the Dark sector also based on an assumed similarity between Normal and Dark sectors implicit in the UST dimension array.

The Layer groups are placed in the four Normal and Dark UST layers. However their four dimension fermion representations distribute the fermions across the four layers, generation by generation, as seen in Figs. 7.3 and 7.5. Thus the Layer groups support inter-layer mixing. The 8 Layer groups are independent of each other.

The Generation groups' four dimension fermion representations treat the fermions within the generations of each layer separately. Figs 7.3 and 7.4 symbolically show the mixing patterns within generations.

The ElectroWeak and Strong groups and interactions mix the fermions within each layer separately. Each layer has its own ElectroWeak and Strong groups. There are eight sets of ElectroWeak and Strong interaction groups in the US: four sets in the Normal sector and four sets in the Dark sector. The ElectroWeak and Strong groups of the Normal and Dark sectors have analogous forms. But they are independent of each other. See Fig. 7.6.

The Connection groups are described in Appendix 7-B. The 7 U(2) Connection groups are independent of each other.

The Cosmos Theory UST *independent* groups (Figs. 7.4, 7.6 and 7-B.2) in the Normal and Dark sectors are:

8 SU(4) Strong interaction Groups
8 SU(2)⊗U(1) ElectroWeak Groups
8 Generation Groups

---

[44] Please note the contents of UST generations differ from the usual form of generations in The Standard Model. See Fig. 7/3 for the form of UST fermion generations and layers.

The Standard Model Generations		
1	2	3
u	c	t
d	s	b
e	μ	τ
ν_e	ν_μ	ν_τ

Generation:

[45] We define a species to be one of the 16 columns in the fermion dimension array.

8 Layer Groups
7 U(2) Connection Groups
1 SL(2, C) Space-time Coordinates

Taking all the UST groups together we see that they implement the principle:

*A fermion in any block has interactions either directly, or indirectly,
with every other fermion in every other block.*

This principle is required for the simple reason that fermions that do not connect to the others are outside of Physics – they have no accessible experimental significance.

7.2 The Species Columns of the UST

The eight lowest mass fermions generation appears in layer 1. See Fig. 7.3. We define a species to be a column with a lowest mass fermion as its identifier at its top (Fig. 7.6.). We designate each species by the lowest mass fermion.

We now consider the form of each species column:

- *In each layer* the fermion quantum numbers are the same with the exception of the Generation group number which we number tentatively 1, 2, 3, 4 for the four generations. See Fig. 7.6 for the Generation group representations of all four layers. Each fermion also has a Layer group number specifying its layer.

- *In each Layer group representation* the species column's Generation group number is the same. All internal symmetry group quantum numbers are also the same except the Layer group number differs within the representation. The Layer group numbers may be taken to be 1, 2, 3, 4. (Fig. 7.5 shows the Layer representations.)

We now assume each of the 16 species columns of Fig. 7.6 has the fermions in all layers perfectly aligned such that *all fermions in a given species column have the same quantum numbers except possibly for one quantum number that labels the generation and one quantum number that labels the layer number of each fermion.*[46]

The reason for this assumption: we require Generation group rotations and Layer group rotations in their representations do not change internal symmetry properties.

We define sixteen species for the Normal and Dark fermions. Figs. 7.4 and 7.6 lists the fermions representative of species: ν, e, q^{up}_1, q^{do}_1, q^{up}_2, q^{do}_2, q^{up}_3, q^{do}_3, ν, e, q^{up}_1, q^{do}_1, q^{up}_2, q^{do}_2, q^{up}_3, q^{do}_3 where q is a quark label.

---

[46] And one quantum number that labels Normal or Dark fermion, the *Darkness number.*

*The 256 masses of the fermions depend on a generation number, a layer number and a quantum number distinguishing Normal from Dark fermions, the **Darkness Number,** plus the usual Strong and ElectroWeak quantum numbers.*

7.3 The Representations of Generation, Layer and Connection Groups in the UST

The fermion fundamental representations of these groups require explanation. The ElectroWeak and Strong interaction groups have representations that span the set of fermions in the first UST layer. The other three layers each have three sets of similar matching ElectroWeak and Strong groups. In total there are eight such groups in the Normal sector and eight such groups in the Dark sector (not counting Generation and Layer groups).. See Fig. 7.1.

7.3.1 Fundamental Generation Representations

There are 64 fermion Generation representations in layer 1 with each consisting of four fermions for the UST in Fig. 7.4. Each layer has a separate Generation group in the Normal and in the Dark sectors. The U(4) Generation group of each layer[47] have a representation for each species in its layer. Thus there are 64 representations for each Generation group of each layer. A Generation group representation supports rotations of its four fermions.

The fermions in each representation in each layer have the same quantum numbers except for the generation number which can range from 1 through 4. The rotations within each species are independent of those in other species and other layers.

7.3.2 Layer Representations

There are 256 Layer group representations = 4 layers × 4 generations × 16 species in the UST. A Layer group representation supports rotations of the fermions in a four dimension representation *of a specified generation number*. Fig. 7.5 displays the Layer representation form for each of the four generations. The Layer groups have a representation for each generation of each species. The rotations within each species are independent of those in other species.

Implicit in the Layer group formulation is the sameness of the Layer representation of the sets of fermions in all layers for internal quantum numbers except for the layer number and Darkness number.

7.3.3 Connection Group Representations

There are 7 U(2) Connection groups in the UST. They are shown in Fig. 7-B.2. Each group has representations connecting subsectors of the 256 fundamental fermion spectrum. In each connection, between two subsectors, the U(2) group connects *corresponding* pairs of fermions. The fermions in a pair have the same quantum numbers except they differ by layer number (vertical lines in Fig. 7-B.2) or the Darkness number (horizontal lines in Fig. 7-B.2).

---

[47] The U(4) group has four diagonal quantum numbers for four baryon number operators.

Implicit in the Connection groups' formulation is the sameness of the representation of the sets of fermions in all layers for internal quantum numbers except for the layer number and/or the Darkness number.

7.3.4 ElectroWeak and Strong Groups Representations

Fig. 7.6 shows horizontal span of the ElectroWeak and Strong groups in each layer. Together with the Generation and Layer groups they cover the complete Normal fermion spectrum and the complete Dark sector independently. They furnish the basis of an extended CKM matrix.

7.4 Conclusion

The introduction of the Generation, Layer and Connection groups by this author imposes a structure on the UST fermion spectrum making each species have the same internal quantum numbers except possibly for Generation number, Layer number, and Darkness number.

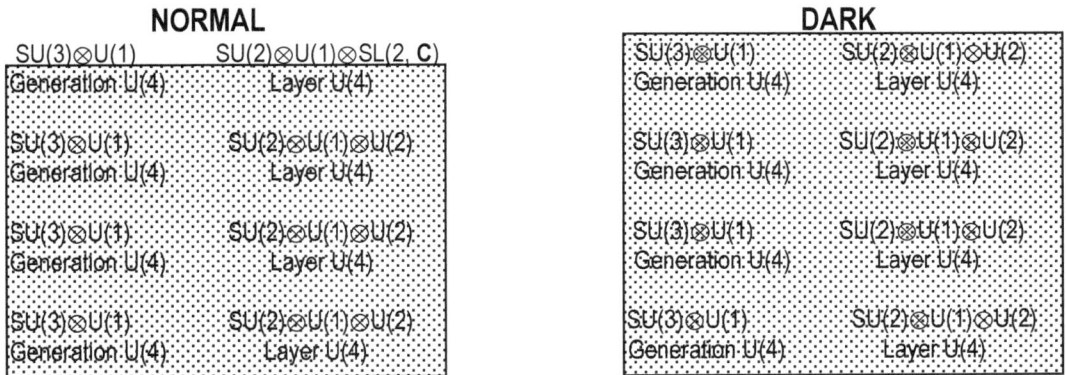

Figure 7.1. The transformed/broken sets of symmetries in the UST. The darkened parts have not as yet been found experimentally. The one undarkened line is for experimentally known groups. Note each item above has an 8 real dimension representation. Note the seven U(2) Connection groups. The SL(2, **C**) representation has four coordinates.[48]

---

[48] The Lorentz Group $SO^+(1, 3)$ is often specified with an SL(2, **C**) representation.

T^1	T^2	T^3	T^4
T^5	T^6	T^7	T^8
T^9	T^{10}	T^{11}	T^{12}
T^{13}	T^{14}	T^{15}	I

Figure 7.2. The dimension array for the 16 U(4) T_k matrices of U(4) and SU(4) interaction terms. Each of 16 submatrices is a 4 × 4 U(4) matrix. The 16th submatrix is the identity matrix.

The Fermion Periodic Table (r = 4)

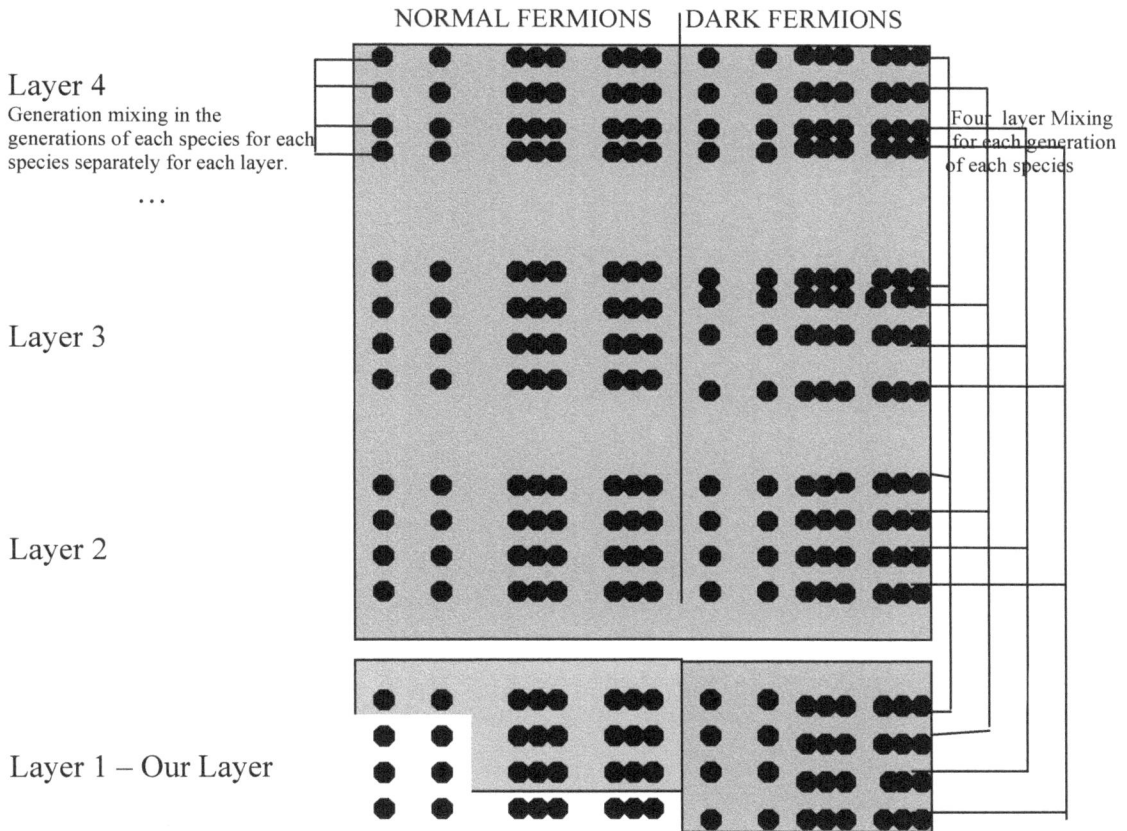

NORMAL FERMIONS DARK FERMIONS

Layer 4

Generation mixing in the
generations of each species for each
species separately for each layer.

. . .

Four layer Mixing
for each generation
of each species

Layer 3

Layer 2

Layer 1 – Our Layer

Figure 7.3. The UST Fermion particle spectrum and partial examples of the pattern of fermion mixing of the Generation groups and of the Layer groups. Unshaded fermion dots are the known fermions, The lines on the right side show Layer group mixing (for Normal and Dark matter) with the mixing among all four layers for each of the four generations individually. There are four Layer groups for Normal matter and four Layer groups for Dark matter. There are 256 fundamental fermions. From Blaha (2018e).

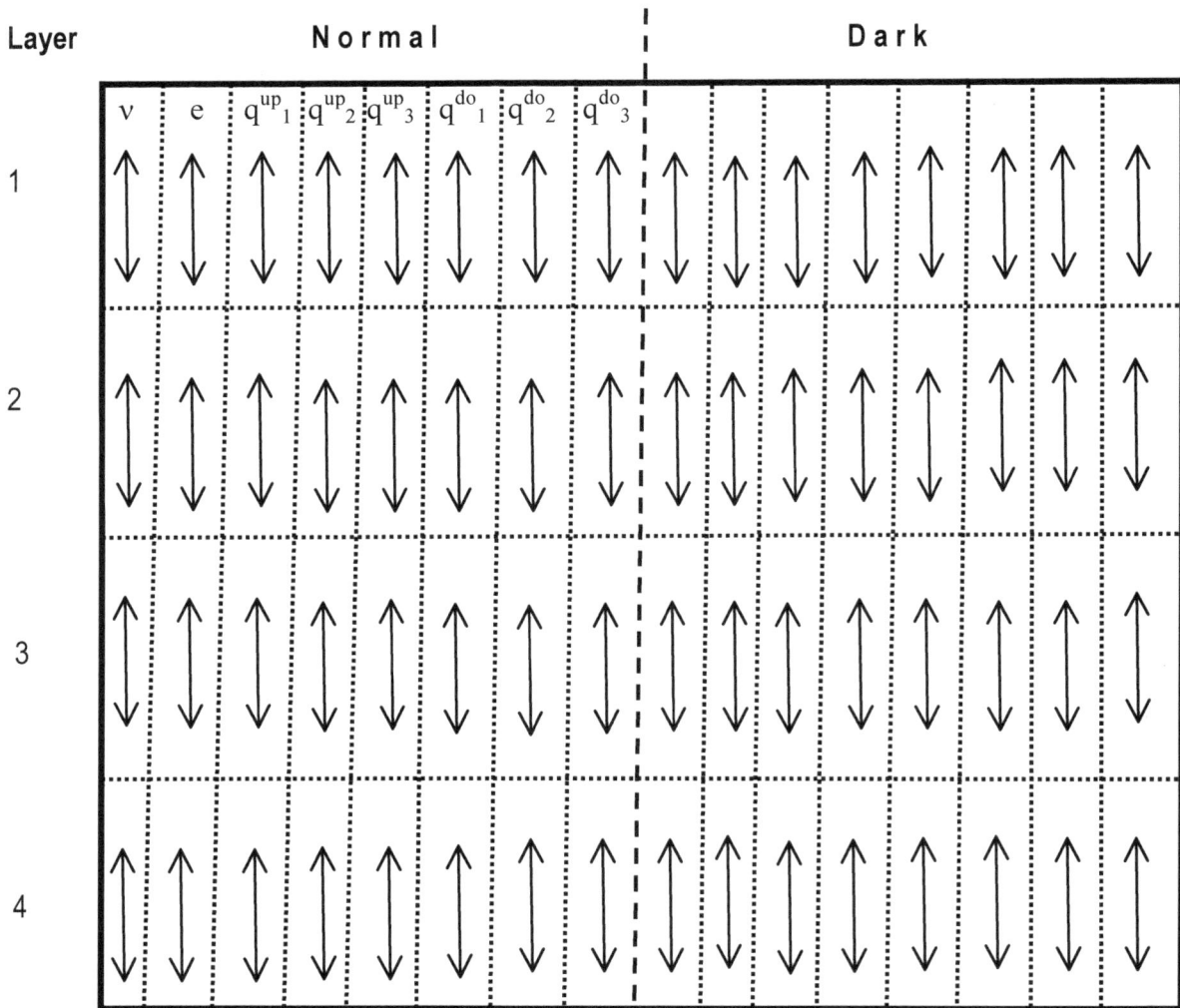

Figure 7.4. Symbolic display of the 8 Generation group representations for fermions of each fermion species within each layer. The "do" superscript indicates "down" type.

The Fermion Periodic Table (r = 4)

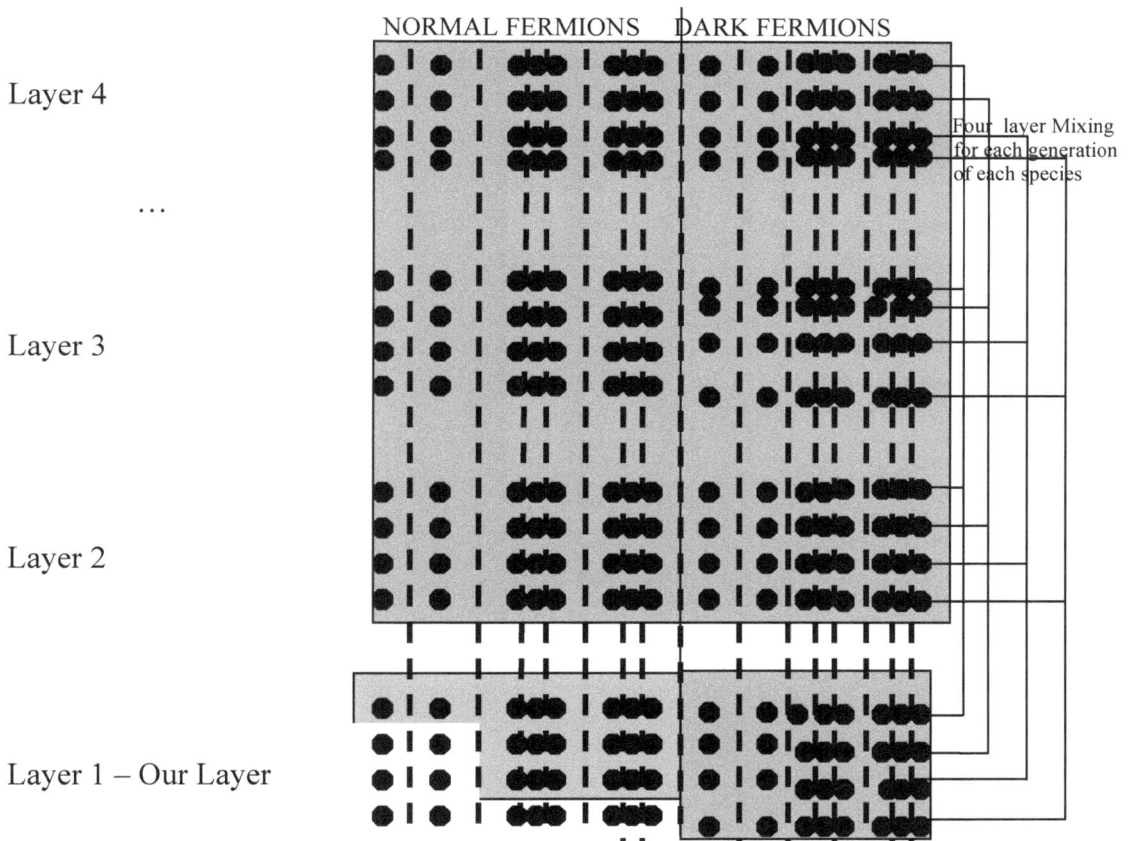

Figure 7.5 Symbolic display of Layer Group representations. Each species has separate representations for each generation. Unshaded parts are the known fermions including μ, v_μ, τ, and v_τ leptons. Shaded parts are yet to be found. From Blaha (2018e).

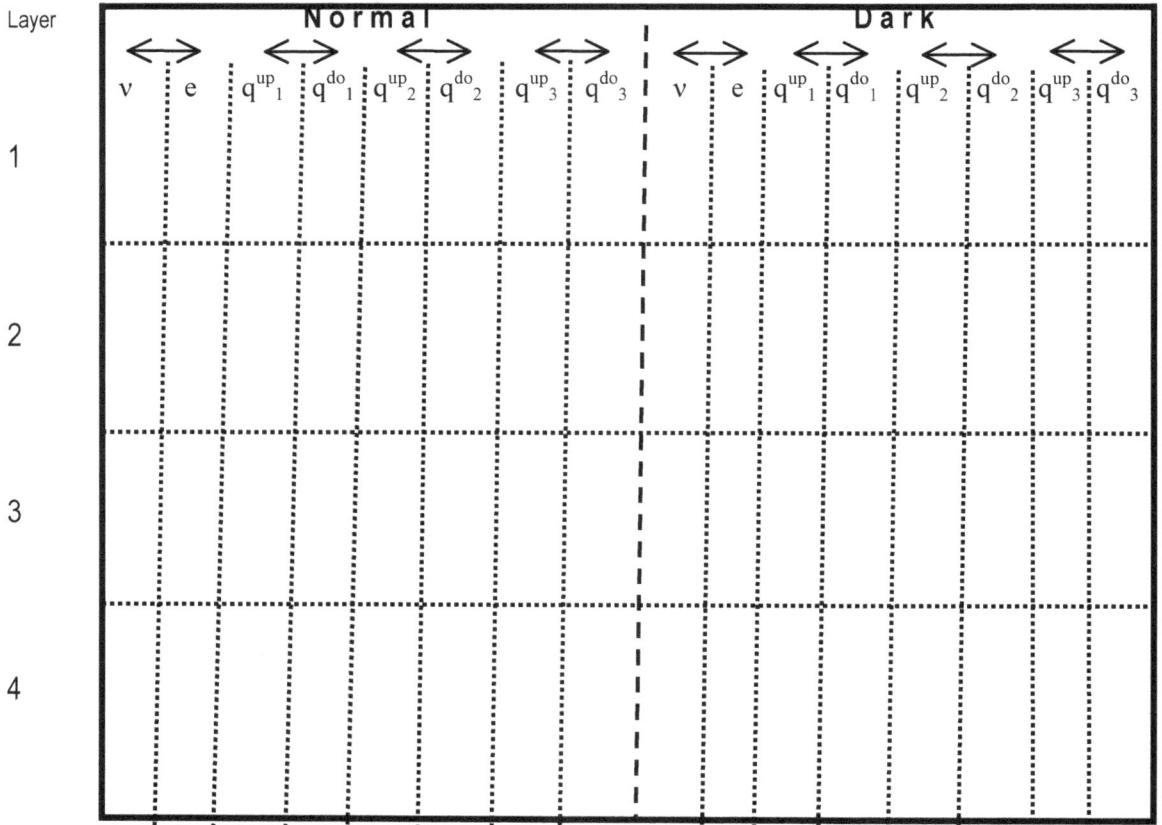

Figure 7.6. ElectroWeak SU(2)⊗U(1) form. Together with SU(4) (not shown) they span an entire layer.

Appendix 7-A. Generation and Layer[49]

7-A.1 U(4) Generation Groups

In the Big Bang all particles were massless and all symmetries unbroken. Hence the four Normal particle number symmetries, and the four Dark particle number symmetries, are all "conserved" in the Big Bang. Afterwards conservation laws are then broken.

We define two particle number operators for normal up-quark particles and down-quark particles, B_{uq} and B_{dq}. Similarly we define two particle number operators for normal species "e" (electron) particles and species "ν" particles, B_e and B_v. Similarly we define Dark matter equivalents:[50] B_{De}, B_{Dv}, B_{Duq}, and B_{Ddq}.

In the absence of symmetry breaking these fermion particle number operators would be conserved. Thus there are two sets of "diagonal" operators with associated U(4) groups for the Normal and Dark sectors. They are part of the Normal U(4) Generation Group and the Dark U(4) Generation Group.

The fermion fundamental representation of a U(4) group has four fermions. U(4) has rotations, and also interactions of the form $\overline{\Psi}\gamma\cdot B\cdot T\Psi$ where Ψ is a fermion four-vector, B is a 16 component U(4) gauge field, and T consists of 16 component 4×4 U(4) arrays.

In the case of the Generation Group the gauge fields have electric charge zero. Since the four species have different electric charges (1, 0, 2/3, -1/3) the U(4) gauge boson fields cannot mix the fermions of different species. Generation Group interactions are diagonal[51] in fermion species (e, ν, up-quark, and down-quark species).

Consequently the U(4) Generation Group for each layer[52] must have a reducible representation D consisting of a set of four fundamental U(4) representations, D_e, D_v, D_{upq}, and D_{dnq}, appearing in blocks along the diagonal of D. Each block is a separate U(4) irreducible representation for a species (due to the electric charge superselection rule.) There is a U(4) Generation group for each of the four layers of the Normal and Dark sectors totaling to 8 generation groups.

There are four generations of each species in the Normal and in the Dark matter sectors. The four generations for each fermion species: e, ν, up-quark, and down-quark each furnish a U(4) fundamental representation within the reducible representation D. The fourth generation of normal fermions has not as yet been found due to their extremely large masses.

---

[49] From Blaha (2017c).,(2018e), (2019g), and (2023d).
[50] By analogy, we assume that there are four species of Dark matter: charged Dark leptons, neutral Dark leptons, Dark up-type quarks, and Dark down-type quarks. Thus we are led to the Dark particle numbers: Dark Baryon Numbers, and Dark Lepton Numbers shown above.
[51] ElectroWeak interactions can cross between species due to their charged gauge vector bosons.
[52] For the Normal and Dark sectors separately.

The Generation Group rotates the fundamental fermions of each fundamental representation separately for each of the four species of each of the four layers.[53] Thus the Generation Group guarantees that all generations of each species have the same electric charge and other quantum numbers.

The U(4) Generation Group also specifies a gauge field interaction among the fermions of its fundamental representation, species by species, for both Normal and Dark sectors. The form of the interactions for the Normal sector for each fermion layer is:

$$g_e \overline{\Psi}_e \gamma \cdot B_e \cdot T\Psi_e + g_v \overline{\Psi}_v \gamma \cdot B_v \cdot T\Psi_v + g_{upq} \overline{\Psi}_{upq} \gamma \cdot B_{upq} \cdot T\Psi_{upq} + g_{dnq} \overline{\Psi}_{dnq} \gamma \cdot B_{dnq} \cdot T\Psi_{dnq}$$

$$(7\text{-}A.1)$$

where g_e ... are coupling constants, the gauge vector fields are B_e ... , and the Ψ_e ... are 4-vectors of fermions of the four generations of each species in a layer.

The gauge vector bosons of the Generation Group have large masses. If the conservation of the fermion particle numbers is broken then we view it as a consequence of Generation Group symmetry breaking.

Generation Group rotations guarantee the internal quantum numbers of each generation of each species are the same since symmetry breakdown is not present at the instant of the Big Bang.

The above discussion applies similarly to the Dark sector. Thus there are 8 Generation Groups in total.

See Blaha (2019g) and (2018e) for a detailed discussion of the Generation Groups.

7-A.2 U(4) Layer Groups

The set[54] of particle number operators can be extended if we take account of the fourfold fermion generations.

We can subdivide the above particle number sets into four additional particle numbers *per generation*. For the ith generation (of the four generations) we define

L_{ie} – The "e" species particle number for the ith generation
L_{iv} – The v species particle number for the ith generation
L_{iuq} – The up-quark species particle number for the ith generation
L_{idq} – The down-quark species particle number for the ith generation

L_{iDe} – The Dark "e" species particle number for the ith generation
L_{iDv} – The Dark v species particle number for the ith generation
L_{iDuq} – The Dark up-quark species particle number for the ith generation
L_{iDdq} – Dark down-quark species particle number for the ith generation

---

[53] There are separate Generation groups for each layer.
[54] Here again, in the Big Bang all particles were massless and all symmetries unbroken. Hence particle numbers are "conserved" in the Big Bang. Conservation is then broken afterwards in most cases.

for each generation i = 1, 2, 3, 4. Individual fermions have positive $L_{ia} = +1$ values and antifermions have negative $L_{ia} = -1$ values for each species.

At this point we have a set of four particle number operators for each of four generations (i = 1, 2, 3, 4) of fermions in the Normal sector and similarly in the Dark sector. We then define a U(4) group framework for each set of particle numbers.

The only way to specify fundamental representations for each of the four sets in a sector is to assume there are four layers, with each layer having four generations, and with a fundamental U(4) representation defined for each generation composed of fermions from each layer. Thus there are four Layer Groups for each Normal and each Dark sector: a Layer Group for generation 1, a Layer Group for generation 2, and so on.

The Layer Groups are also "split" by species due to the electric charge superselection rule. Each Layer Group is diagonal in the four fermion species. All their gauge fields are electrically neutral. Thus the Layer Group fundamental representations of the Normal sector total 16.[55] The Layer Group fundamental representations of the Dark sector also total 16. There are 4 Normal Layer Groups and 4 Dark Layer Groups.

Consequently each of the four U(4) Layer Groups in the Normal fermion sector has a reducible U(4) representation D_j for j = 1, 2, 3, 4. Each reducible representation is composed of four irreducible U(4) representations for each species due to the electric charge superselection rule:

$$D_j = D_{je} + D_{jv} + D_{jupq} + D_{jdnq},$$

for j = 1, 2, 3, 4.

There are four layers of each species in the Normal and in the Dark matter sectors. The second, third and fourth layers of normal fermions has not as yet been found due to their extremely large masses.

A Layer Group rotates the fundamental fermions of each fundamental representation separately for each of the four species of each of the four generations.

The Layer Groups guarantee that all layers of each species have the same electric charge and other quantum numbers.

Each U(4) Layer Group also specifies a gauge field interaction among the fermions of its fundamental representation, species by species, for both Normal and Dark sectors. The form of the interactions is:

$$g_{ei}\overline{\Psi}_{ei}\gamma \cdot C_{ei} \cdot T\Psi_{ei} + g_{vi}\overline{\Psi}_{vi}\gamma \cdot C_{vi} \cdot T\Psi_{vi} + g_{upqi}\overline{\Psi}_{upqi}\gamma \cdot C_{upqi} \cdot T\Psi_{upqi} + g_{dnqi}\overline{\Psi}_{dnqi}\gamma \cdot C_{dnqi} \cdot T\Psi_{dnqi}$$

$$(7\text{-}A.2)$$

for i = generation = 1, ... , 4, where g_{ei} ... are coupling constants, the gauge fields are C_{ei} ... , and the Ψ_{ei} ... are 4-vectors of fermions formed of the ith generation fermions in each layer of each species.

The gauge vector bosons of the Layer Groups also have large masses. If the conservation of the fermion particle numbers is broken then we view it as a consequence of Layer Groups symmetry breaking.

---

[55] Four Layers groups irreducible representations for the generations × four species = 16 irreducible representations.

Layer Group rotations guarantee the internal quantum numbers of each layer of each species are the same since symmetry breakdown is not present at the instant of the Big Bang.

The above discussion applies similarly to the Dark sector. There are 16 Dark Layer Groups.

Fig. 7.3 shows the fundamental fermion spectrum with the representations of the Generation groups and Layer groups indicated.

Experimentally, we know of three generations of fermions—the lowest 3 generations of the lowest level. The remaining 4th generation and three layers of fermions are of much higher mass and are yet to be found.

See Blaha (2019g) and (2018e) for a detailed discussion of the Layer Groups. We note in passing that the symmetries of these number operators are badly broken. Yet the underlying group structure remains.

Appendix 7-B. Connection Group Symmetries and HyperCosmos Space-Time Coordinates

Hypercomplex numbers are known to be related to symmetry groups. In this section we consider the multiple space-time symmetries that appear in the separation of dimension arrays into representations of groups. We have suggested in Blaha (2021b), (2021e) and (2021g) that the set of hypercomplex space-time dimensions be transformed to a set symmetry groups in each HyperCosmos space.

There is a two-fold justification for this procedure: 1). There is no evidence for the existence of hypercomplex numbers in our universe. 2) The generated set of symmetry groups has the important purpose of providing ultra-weak interactions uniting Normal and Dark matter. Without unification, Dark matter becomes physically irrelevant for elementary particle physics.

We begin by noting the space-time dimension is set by the dimension array:

$$r = \log_2 (d_{dN}/16) \qquad (7\text{-}B.1)$$

For N = 9 the space-time dimension is 0 since d_{dN} = 16. For N = 7 (our universe's space) the space-time dimension is 4 since d_{dN} = 256. We will consider the cases of the UST, Megaverse, and Maxiverse to illustrate the method.

7-B.1 Hypercomplex Coordinates Transformed to Symmetry Groups in Our Universe N = 7

Our UST formulation for our N = 7 universe has a d_{d7} = 256 component dimension array. A preliminary view of the symmetry groups that it implies appears in Fig. 7-B.1. Note that there are 32 dimensions that are initially allocated to space-time dimensions suggesting a hypercomplex space-time. We chose to allocate 4 real dimensions to obtain the 4-dimension space-time implied by eq. 7-B.1. The remaining 28 real dimensions we allocated to additional symmetry groups—namely seven U(2) groups. We call these groups Connection groups since their role is to "connect" fermions residing in different fermion spectrum blocks.

The structure of the seven additional U(2) groups is not specified. We chose to use a reasonable physical principle to allocate them. We believe their role is to "connect" fermions in different blocks since the fermions within each fermion block have "known" interactions. Note that there are initially eight blocks, each with their own set of symmetry groups and corresponding interactions, and initially no interactions between the eight blocks. If there are no block interactions (except gravity), then the Physics of the fermion set is conceptually disjoint. In the absence of interactions the disjoint part is irrelevant – not physical. Thus we choose to implement inter-block interactions as in Fig. 7-B.2 following the principle:

A fermion in any block has interactions either directly, or indirectly,
with every other fermion in every other block.

Fig. 7-B.2 shows an implementation of this principle. The horizontal lines in Fig. 7-B.2 indicate 1:1 transformations between all corresponding fermions of each "Normal" and "Dark" block. The three "angled" lines indicate 1:1 transformations between corresponding fermions of a "Normal" and a "Dark" fermion block in the layer above it. The result is a see-saw type of pattern.

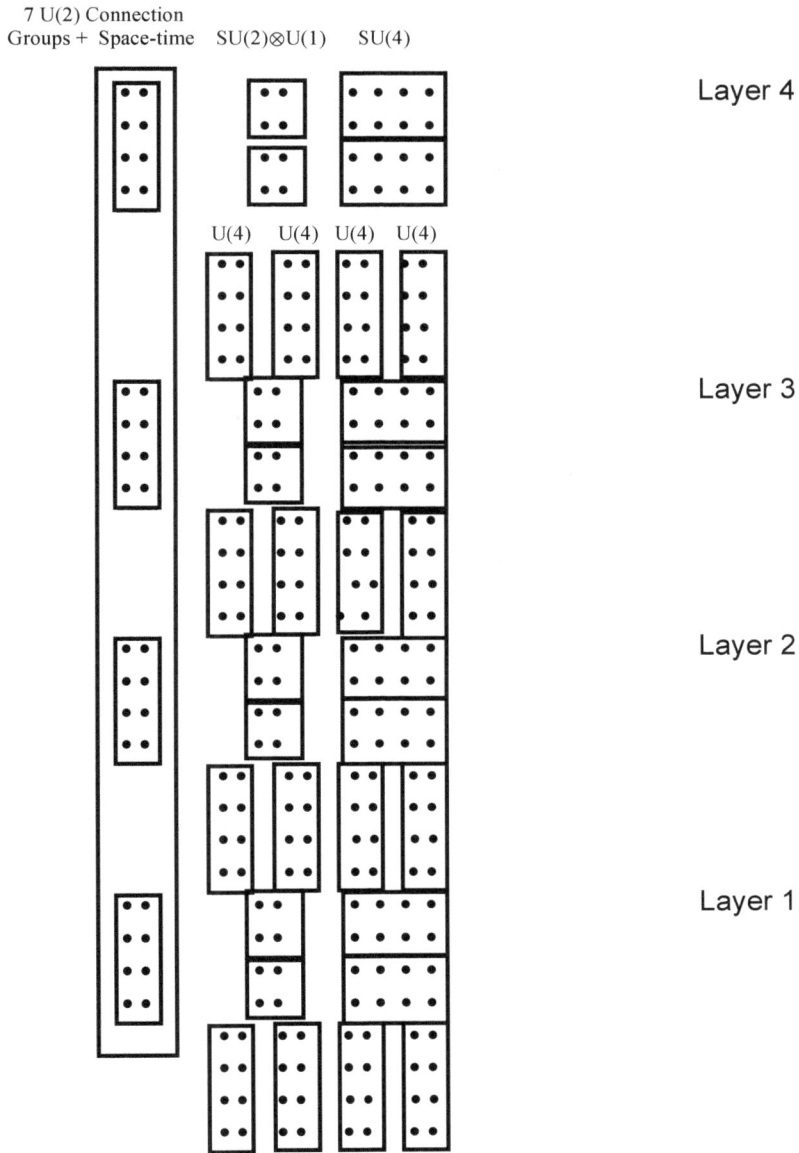

Figure 7-B.1. The four UST layers internal symmetry groups (and space-time) with SU(4) before breakdown to SU(3)⊗U(1). Note the left column of blocks are combined below to specify a 4 dimension real space-time plus seven U(2) Connection groups. Note each layer has 64 dimensions = 56 + 8 dimensions.

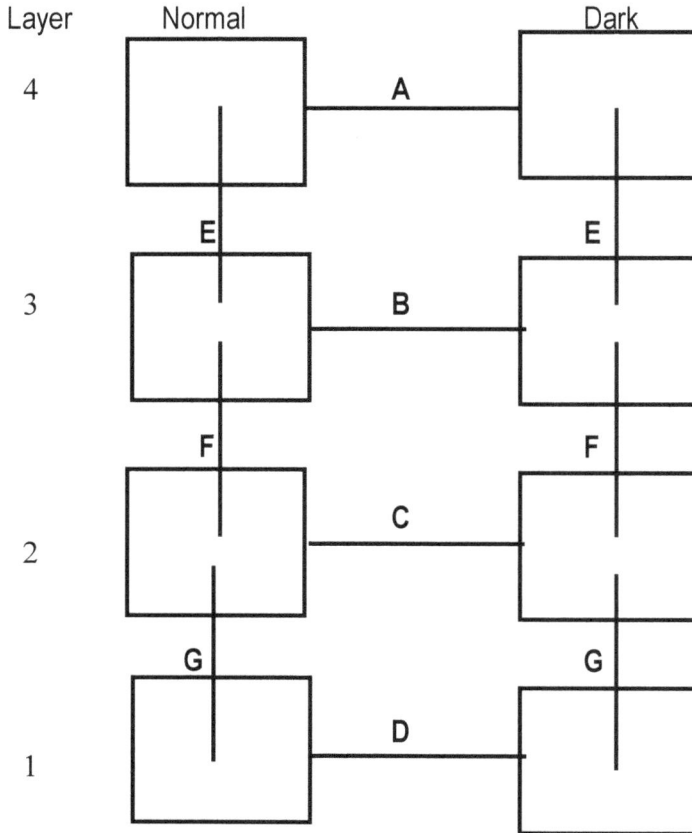

Figure 7-B.2. The seven U(2) Connection groups (shown as 10 lines) between the eight UST blocks. Connection groups are obtained by transfering 28 dimensions from UST space-time to internal symmetries with the consequent reduction of the space-time from four octonion (complex quaternion) coordinates to four real coordinates. The Connection groups generate rotations and interactions between corresponding fermions and vector bosons of each pair of blocks. The Normal and Dark sector U(2) vertical connections above (E, F, G) represent the same U(2) groups.

7-B.1.1 The U(2) Connection Groups

The seven U(2) Connection groups of Fig. 7-B.2 generate "rotations" and interactions between *corresponding* fermions and vector bosons of each pair of blocks of the eight blocks of fermions in the UST.

7-B.1.1.1 Horizontal Lines

The horizontal lines in Fig. 7-B.2 (A, B, C, and D) each represent a U(2) Connection group that "rotates" two *corresponding* fermions in the Normal and Dark sectors of each layer. Thus a Normal e is "rotated" with a corresponding Dark e, and so on.

Each of the four horizontal Connection Groups has a reducible U(2) representation D that is the sum[56] of 4*8 = 32 irreducible U(2) representations. We may view the irreducible representations D_j as strung along the diagonal.

$$D = \sum_{j=1}^{32} D_j \qquad (7\text{-B.}2)$$

for each of the U(2) groups of the four horizontal lines in Fig. 7-B.2.

The U(2) group also specifies gauge field interactions between corresponding fermions in each layer of the Normal and Dark sectors of the form

$$g\overline{\Psi}_{Nn}\gamma\cdot A\cdot T\Psi_{Dn} \qquad (7\text{-B.}3)$$

where N indicates a Normal fermion and D indicates the corresponding Dark fermion, with A being a U(2) gauge vector boson, and n the label for corresponding fermions.

These U(2) transformations imply that the Normal and Dark sectors have the same species.

7-B.1.1.2 Vertical Lines

The pairs of vertical lines in Fig. 7-B.2 (E, F, G) each represent a U(2) Connection group that "rotates" sets of two *corresponding* fermions in adjacent layers as shown in Fig. 7-B.2 in the Normal and Dark sectors. Thus a Normal e in layer 1 is rotated with a corresponding Normal e in layer 2, and so on.

Each of the three (six counting both Normal and Dark lines in Fig. 7-B.2) vertical Connection Groups has a reducible U(2) representation D that is the sum of 64 irreducible U(2) representations.[57] We may view D as an array of 64 U(2) irreducible representation dimensions D_j strung along the diagonal.

$$D = \sum_{j=1}^{64} D_j \qquad (7\text{-B.}4)$$

for each of the U(2) groups of the 3 (6) horizontal lines in Fig. 7-B.2. Note the 64 irreducible representations include both Normal and Dark sectors of a layer. [58]

Each U(2) group also specifies a gauge field interaction between corresponding fermions in adjacent layers for both Normal and Dark sectors:

---

[56] Eight fermions per generation ▪ four generations, thus accounting for each fermion in a block.
[57] Eight fermions per generations ▪ four generations ▪ 2 types of matter (Normal and Dark).
[58] There are 64 fermions in total for each of the four layers of UST.

$$g\overline{\Psi}_{nl_1}\gamma\cdot A\cdot T\Psi_{nl_2} \tag{7-B.5}$$

where l_1 and l_2 designate layers, A is a gauge field vector boson, and n the label for corresponding fermions.

Each E, F, and G U(2) group reducible representation includes both Normal and Dark sectors.

7-B.1.2 The Connection Groups are UltraWeak Interactions

Since there is no convincing experimental evidence for particle interactions between Normal and Dark matter, or between Normal fermion layers the Connection groups appear to be UltraWeak.

7-B.2 MEGAVERSE with Six Real Space-Time Coordinates (Dimensions)

The MEGAVERSE space is a Blaha number N = 6 HyperCosmos space. It is the Megaverse (Multiverse) with 6 real-valued space-time coordinates by eq.7-B.1 above. It has a d_{d6} = 1024 dimension array.

We view the MEGAVERSE dimension array as composed of four copies of the N = 7 dimension array. Initially $4\cdot32 = 128$ dimensions are for space-time coordinates.

We allocate $4\cdot28 = 112$ dimensions to each of the four UST copies within MEGAVERSE to give each copy 7 SU(2) Connection groups. Therefore $4* 28 = 112$ of the 128 space-time dimensions of MEGAVERSE are mapped. The remaining 16 dimensions give 6 real-valued space-time dimensions to the Megaverse space, 8 dimensions to a Megaverse SU(4) Connection group, and 2 dimensions to a U(1) group for all fermions in the Megaverse. See Figs. 7-B.3 and 7-B.4.

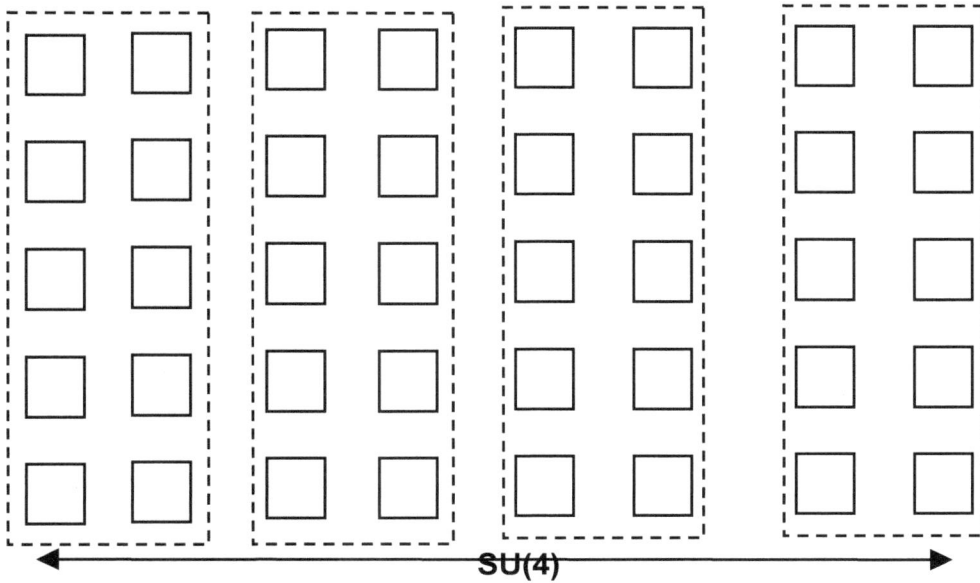

Figure 7-B.3. MEGAVERSE has four UST copies. An SU(4) internal symmetry Connection group maps between corresponding fermions in the four copies: fermion by fermion. An additional U(1) Connection group applies to every corresponding fermion. It is not shown in this figure.

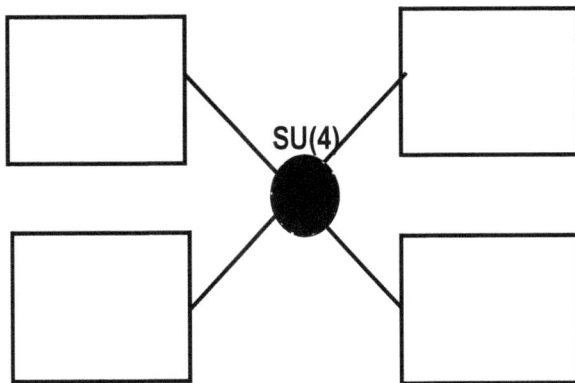

Figure 7-B.4. The SU(4) Connection Group of MEGAVERSE connecting fermions in the four UST "copies" blocks. An additional U(1) Connection group applies to every corresponding fermion. It is not shown in this figure.

7-B.3 Maxiverse with Eight Real Coordinates (Dimensions)

The Maxiverse is a Blaha number $N = 5$ space with a dimension array of $d_{d5} = 4096$ dimensions.

The Maxiverse contains four copies of MEGAVERSE. Each MEGAVERSE has $4*6 = 24$ space-time dimensions. (The remainder in each MEGAVERSE copy consists of internal symmetries and Connection groups.) We allocate the 24 dimensions to eight real space-time dimensions plus 16 dimensions for a new ultraweak (possibly broken) SU(8) Connection group for the four parts of the Maxiverse. See Fig. 7-B.5.

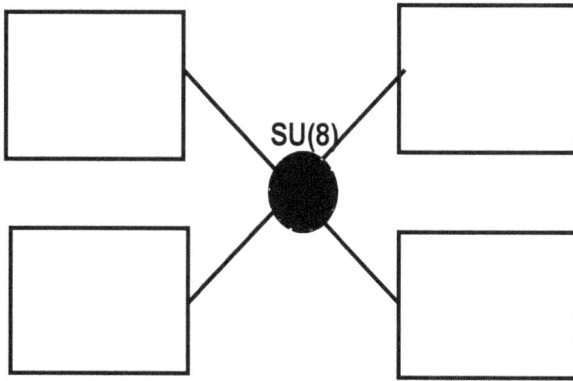

Figure 7-B.5. The SU(8) Connection Group of Maxiverse connecting fermions in the four MEGAVERSE "copies" blocks.

7-B.4 Determining the Connection Groups for a Space

In the previous sections we determined the Connection Groups for $N = 7$, $N = 6$, and $N = 5$ spaces. The dimensions required for the Connection Groups of a space d_{gN} for $N \leq 6$ can be obtained by noting the space-time dimension of a space-time r is related to the space-time dimension $r - 2$ of the next lower space. Since a space is four copies of the next lower space it has $4(r - 2)$ space-time dimensions from the four duplicates. Then, using eq. 2.17, we find

$$4r_{N+1} = 4(r_N - 2) = r_N + d_{gN}$$

or

$$d_{gN} = 3r_N - 8 \tag{7-B.6}$$

where d_{gN} is the *real-valued* space-time dimensions reallocated to Connection Groups in space N. For $N = 6$, $r_6 = 6$, $d_{gN} = 10$. For $N = 5$, $r_5 = 8$, $d_{gN} = 16$. And so on.

Eq. 7-B.6 determines the dimensions available for Connection Groups for the space with Blaha number $N \leq 6$. Using eq. 7-B.1:

$$r_N = \log_2 (d_{dN}/16) \tag{7-B.7}$$

we relate d_{gN} to the size of the dimension array:

$$d_{gN} = 3 \log_2(d_{dN}/16) - 8 \qquad (7\text{-}B.8)$$

thus specifying the number of space-time dimensions allocated to space number N.

8. Periodic Table of UST Fermions

The Periodic Table of Chemical Elements categorizes elements by column (groups) and rows (periods). In addition it identifies some sets of elements, such as metals and actinides, as blocks.

In this chapter we will consider the Periodic Table of Fermions. They are the analog of the Chemical Periodic Table.

We will show that the UST fundamental fermions may be viewed as a periodic table with its parts characterized by symmetry groups. The form of this periodic table is fairly sophisticated because of the nature of the Generation, Layer, and Connection groups. We will characterize the parts of the Fermion Periodic Table by the symmetry groups that govern it. There are five types of internal symmetry groups with interactions: Strong, ElectroWeak, Generation, Layer and Connection groups.

In describing the role of the various interactions in the Fermion Periodic Table we will consider fermion representations that are patterned after the form of the fundamental representations of the groups. Thus fermion wave functions will individually be tailored for symmetry group representations. We will treat the Dark sector interactions and representations as the mirror image of their corresponding Normal sector equivalents with one exception: some Connection group interactions connect the Normal and Dark sectors and thus must be considered for the entire set of fermions.

8.1 Fermion Dirac Equation

In this section we will describe the Dirac equation Lagrangian terms for the 256 fundamental fermions of the UST which has an $r = 4$ Cosmos Theory space. We use the PseudoQuantum Quadplex formalism:[59]

$$\mathscr{L} = \overline{\Psi}_1\{\ i\ ^y\gamma^\mu\ \partial/\partial y^\mu + {}^z\gamma^\mu\ \partial/\partial z^\mu + i\ ^u\gamma^\mu\ \partial/\partial u^\mu + {}^v\gamma^\mu\ \partial/\partial v^\mu + \sum_{k=1}^{8} [g_{sk}\ A_{kS}{}^{a\mu}(y)\ ^y\gamma_\mu T_{kSa} +$$

$$+\ ig_{EWk}\mathbf{t}_k(S_z)W_k{}^\mu(y) + ig_{EWk}'t_{k0}(S_z)W_{k0}{}^\mu(y)\ +\ g_{Gk}\ A_{kG}{}^{a\mu}(y)\ ^y\gamma_\mu S_{kGa}\ +$$

$$+\ g_{Lk}\ A_{kL}{}^{a\mu}(y)\ ^y\gamma_\mu S_{La}] + \sum_{k=1}^{7} g_{Ck}\ A_{kC}{}^{a\mu}(y)\ ^y\gamma_\mu S_{kLCa} + M\}\psi_2(y,\,z,\,u,\,v)\ +$$

$$+\ \overline{\Psi}_2\{\ i\ ^y\gamma^\mu\ \partial/\partial y^\mu + {}^z\gamma^\mu\ \partial/\partial z^\mu + i\ ^u\gamma^\mu\ \partial/\partial u^\mu + {}^v\gamma^\mu\ \partial/\partial v^\mu + \sum_{k=1}^{8}[g_{sk}\ A_{kS}{}^{a\mu}(y)\ ^y\gamma_\mu T_{kSa} +$$

$$+\ ig_{EWk}\mathbf{t}_k(S_z)W_k{}^\mu(y) + ig_{EWk}'t_{k0}(S_z)W_{k0}{}^\mu(y)\ +\ g_{Gk}\ A_{kG}{}^{a\mu}(y)\ ^y\gamma_\mu S_{kGa}\ +$$

$$+\ g_{Lk}\ A_{kL}{}^{a\mu}(y)\ ^y\gamma_\mu S_{La}] + \sum_{k=1}^{7} g_{Ck}\ A_{kC}{}^{a\mu}(y)\ ^y\gamma_\mu S_{kLCa} + M\}\psi_1(y,\,z,\,u,\,v) \qquad (8.1)$$

---

[59] Each set of coordinates (y, z, u, v) should use Two Tier coordinates in the evaluation of perturbation theory diagrams to guarantee finite results to any order in perturbation theory.

where the LAB space-time is the y coordinate system, and using the Generation and Layer terms:

$$g_G \, A_G{}^{a\mu}(y) \, {}^y\gamma_\mu S_{Ga} + g_L \, A_L{}^{a\mu}(y) \, {}^y\gamma_\mu S_{La} \qquad (8.2)$$

for the eight Generation and eight Layer groups of the Normal and Dark sectors.

The summations are over all the groups. The 256 component wave function has parts for the Normal and Dark fermions. Within each of the parts, which we assume to have the same form, there are sub-parts for each group representation.

8.2 General Form of Wave Function

The general form of the 256 component wave function is:

$$
\Psi =
\begin{bmatrix}
\underline{\textbf{Normal}} & \begin{array}{l} \text{Strong Reps} \\ \text{ElectroWeak Reps} \\ \text{Generation Reps} \\ \text{Layer Reps} \\ \text{Connection Reps} \end{array} \\
\\
\underline{\textbf{Dark}} & \begin{array}{l} \text{Strong Reps} \\ \text{ElectroWeak Reps} \\ \text{Generation Reps} \\ \text{Layer Reps} \\ \text{Connection Reps} \end{array}
\end{bmatrix}
\qquad (8.3)
$$

One layer examples appear in eqs. 6.4 – 6.7.

Below we extend the description *only* to the Normal part of the wave function fermion components for all the interactions except for those of the Connection groups. The Connection groups, in part, connect the Normal and Dark sectors. Their description necessarily involves both Normal and Dark.

We will treat each interaction separately – adapting the description of the wave function to be "diagonal" in its fundamental group representations. The need for this approach is signaled by comparing the Strong and ElectroWeak fermion representation forms in eqs. 6.4 – 6.7. This approach avoids "spaghetti" specifications of interaction group fundamental representations.

For all interactions except the Connection group cases we will only consider the Normal fermion sector. We assume the form of the Dark sector is its "mirror" image.

The following sections, except for the Connections groups section, will consider the Normal fermion sector only.

8.3 SU(4) Interactions and Representations

Different SU(4) groups appear in each Normal layer. We can adapt the order of the components of the overall fermion wave function to the block diagonal ordering of four dimension SU(4) fundamental representations. These representations have 4×4 matrices. We use sequences' blocks of four fermions as the components of the representations. (There are two sequences for each layer as seen earlier.) For one layer we can place the ordering of fermion components of the wave function in the form seen in Fig. 6.5. For the four Normal fermion layers of the UST we obtain the representations of Fig. 8.1.

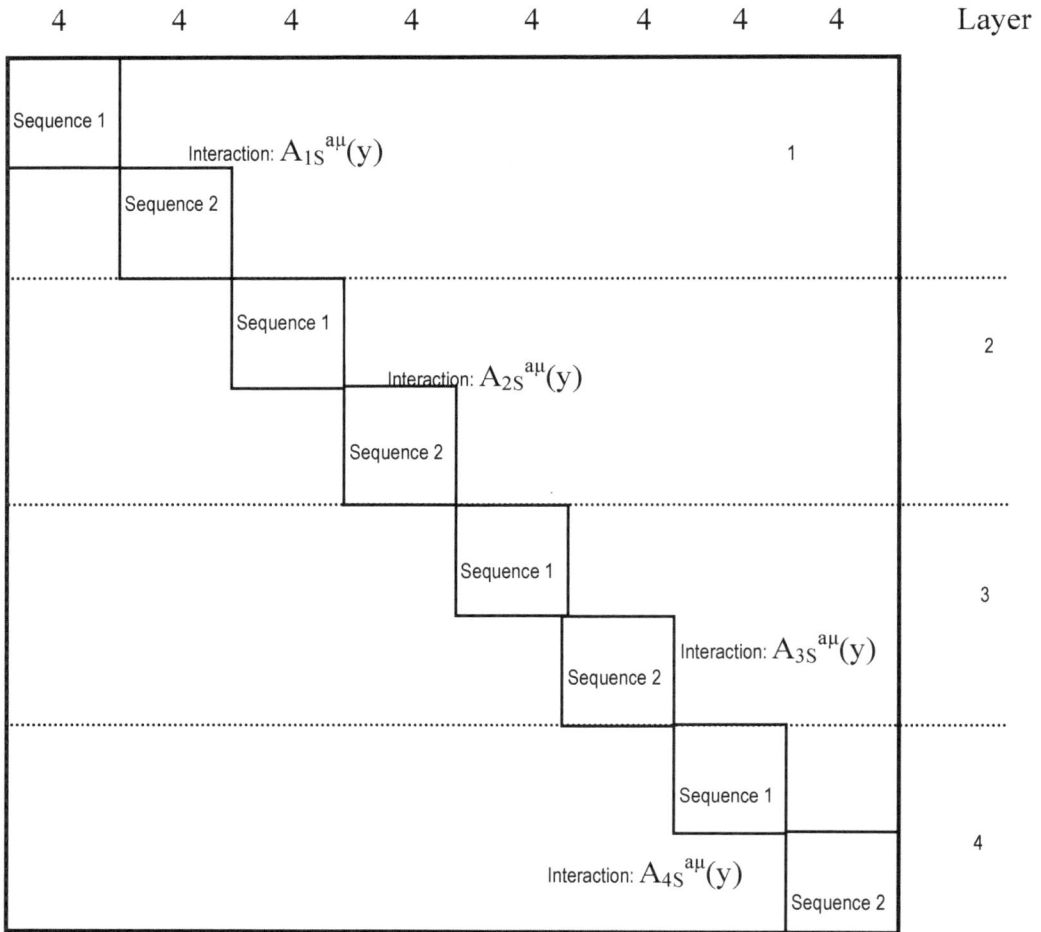

Figure 8.1. Pairs of SU(4) representations for the two sequences of fermions appear in each of the four layers. Each layer has a different SU(4) interaction group.

8.4 SU(2)⊗U(1) ElectroWeak Groups

There is one ElectroWeak group in each layer of the Normal sector. Each layer has 4 pairs of ElectroWeak group representations *for each generation* in the layer. Thus there are 64 ElectroWeak group representations in the 4 layers of the Normal sector. We thus adapt the order of the fermions to the ElectroWeak group representations in the fermion wave function to avoid "spaghetti" organization of representations. We obtain Fig. 8.2. In this figure we use the one generation, one layer view of Fig. 6.7 as a building block using the symbol ■ to represent the entire matrix of ElectroWeak representations of Fig. 6.7.

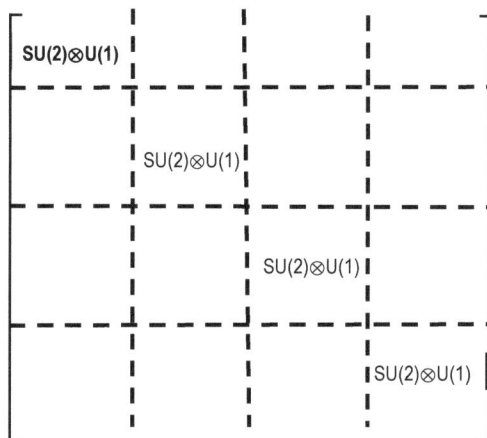

Figure 6.7. Four ElectroWeak representations for one generation of one layer. The representations correspond with the order of the fermions in the fermion wave function vector of Fig. 6.6.

Within the wave function each generation of a layer has the fermion order:

$$\Psi_{EW} = \begin{bmatrix} v' \\ e \\ d \\ u \\ s \\ c \\ b \\ t \end{bmatrix}$$

Figure 6.6. Fermion wave functions terms ordered as four sets of fermion ElectroWeak representations using pairs of fermions in Fig. 6.7.

Layer

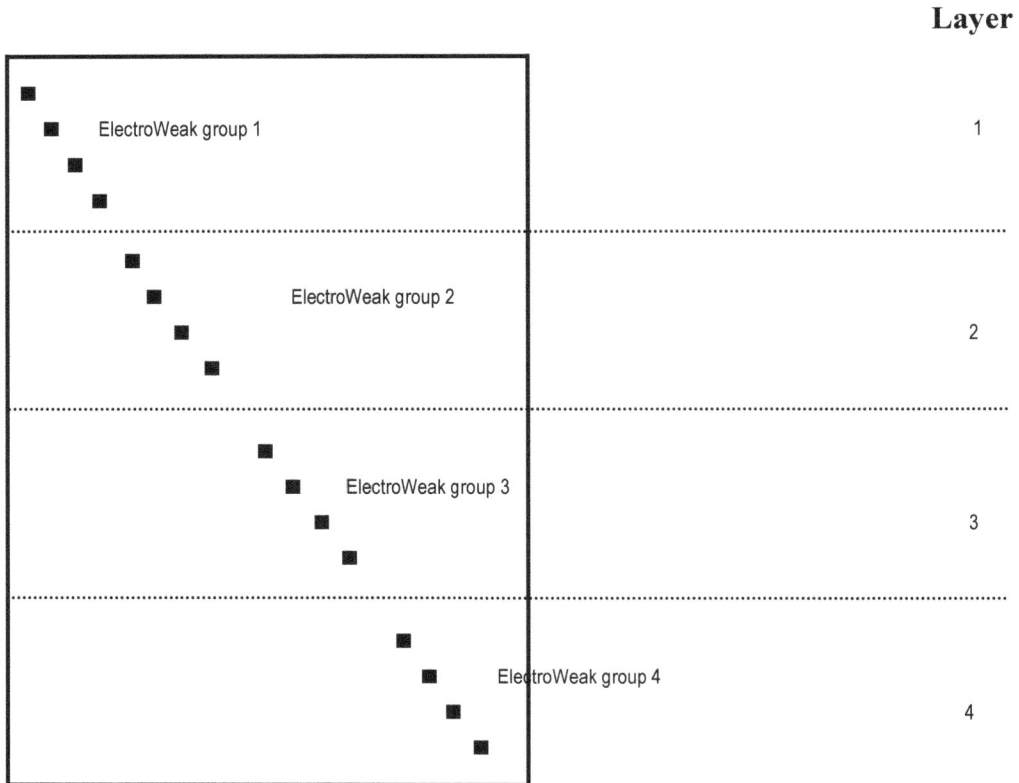

Figure 8.2. Four layers of ElectroWeak representations. Each layer has a set of sixteen ElectroWeak representations. Each ■ corresponds to a one generation, one layer set of four ElectroWeak representations as in Fig. 6.7 above. This figure has 64 Normal ElectroWeak representations in total. Their total number of dimensions is 128 = 2×64 – corresponding to the size of the UST Normal sector. (The complete specification has a size of 256 if one takes account of the Dark sector as expected in the UST.) Note each layer has a different ElectroWeak group and interaction, and 16 representations.

8.5 Generation Groups

The Generation group has 32 representations in the Normal sector of the fermion wave function. There are 8 Normal representations for each of the four layers – one four dimension representation for each species. They appear in Fig. 7.4. If we adapt the wave function vector to the order of fermions in the Generation representations we obtain Fig. 8.3.

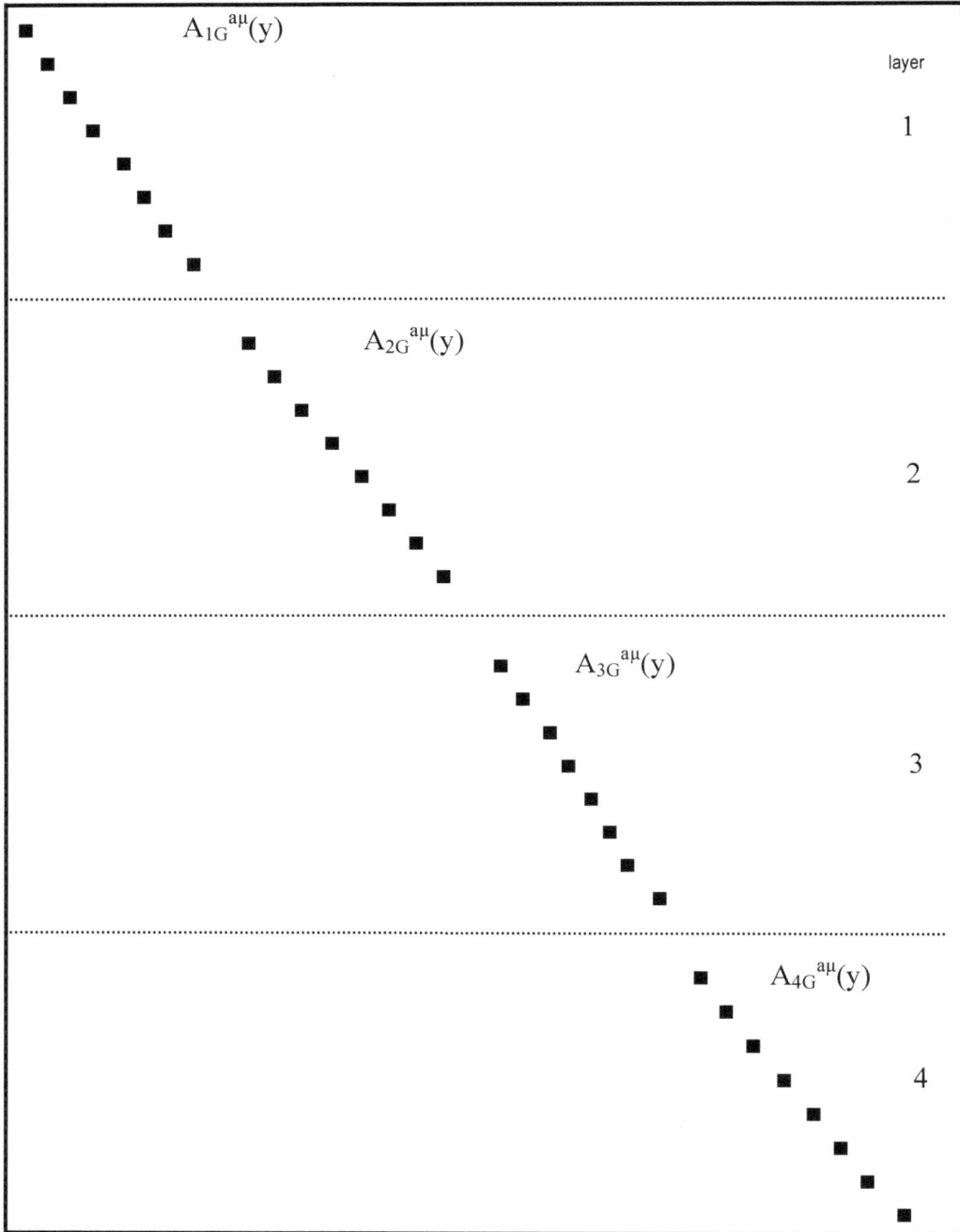

Figure 8.3. Eight Generation group representations for each layer. We specify each four dimension Generation representation with the ■ symbol. The representations are displayed in block diagonal form. In each layer the eight fermion Generation representations are for the species. Each layer has a different Generation group.

8.6 Layer Groups

There are four Layer groups in the Normal part of the UST. We placed these groups in each of the four layers (Fig. 3.6) although each Layer group representation contains a fermion of each layer. See Fig. 7.5. The Layer group representations straddle the layers. So they are best ordered by generations. There is a Layer group representation containing generation 1 of all layers. There is a Layer group representation containing generation 2 of all layers and so on. There is a separate representation for each of the 8 Normal species. As a result there are 32 Layer representations = 4 generations × 8 species. The total number of Normal dimensions of these representations is 128.

9.4 Contents of Each FRF in 4 Dimensions

A HyperCosmos space FRF, before internal symmetries are introduced, has $2^{r/2+2}$ non-zero elements where r is the space-time dimension.[67] The number of non-zero elements equals the number of b's and d's, 16, before internal symmetries are introduced as in eq. 9.1.

9.4.1 Rotations Induced by GR Transformations

These b's and d's are rotated when the wave function is transformed to accelerating coordinates or undergoes a GR transformation. See S. Blaha, "The Local Definition of Asymptotic Particle States", IL Nuovo Cimento **49A**, 35 (1979) for the description of the impact of such transformations.

9.4.1 Rotations Induced by GR and Internal Symmetry Transformations

After internal symmetries are introduced into the b's and d's as in eq. 9.3, the number of b's and d's is 2^{r+4} due to a factor of $2^{r/2+2}$ introduced for internal symmetries. Correspondingly, the number of elements in the dimension array is now 2^{r+4}.

Earlier we found 16 b's and d's in our universe's space – not counting internal symmetries. These b's and d's become the non-zero elements in our universe's FRF. They are subject to GR transformations. If we map them to an FRF they become $2^{r/2+2}$ non-zero entries in the FRF vector. The vector has 2^{r+4} components. The remaining FRF components are $2^{r+4} - 2^{r/2+2}$ zeroes.

These components become non-zero under GR – Internal Symmetry transformations. Then the FRF has $2^{r/2-2}$ replicates of the $2^{r/2+2} = 16$ initial non-zero elements in the space's FRF. In the 4 dimension space of our universe there are initially 240 zeroes and 16 non-zero items in the FRF vector.

Justification

Each HyperCosmos space has a lowest spin fermion wave function with $2^{r/2-2}$ replicates of the b's and d's of eq. 9.1. The replication of b's and d's matches the replication of 16 non-zero element sets in the initial FRF.

The contents of the FRF's are listed in Figs. 9.2 – 9.4 below in terms of Blaha number N. Fig. 9.7 – 9.10 shows the fermion and internal symmetry group content of the r = 4 (N = 7) FRF.

The non-zero elements in an FRF, after it is filled due to a transformation, map to fundamental fermions, Internal Symmetry groups, or dimensions as the case may be.

---

[67] The total number of components in the FRF vector is 2^{r+4} counting non-zero and zero-valued components.

8.7 Connection Groups

There are seven U(2) Connection groups. See Fig. 7-B.2. Three of the groups connecting layers are "vertical." They are labeled E, F, and G. We take them to be the same for both Normal and Dark matter sectors.

8.7.1 Vertical Connection Group Representations

These group representations contain a pair of corresponding fermions: from adjoining layers. The fermion pair has the same quantum numbers except for the Layer group quantum number which differs by one unit. Thus Connection group representations are diagonal in the other quantum numbers. For each Connection group there are 32 U(2) representations. Thus the connection between layer 1 and layer 2 has 32 U(2) representations. The connection between layer 2 and layer 3 has 32 U(2) representations. And the connection between layer 3 and layer 4 has 32 U(2) representations. The total number of "vertical" Connection group representations is 96 in the Normal sector and 96 in the Dark sector.

8.7.2 "Horizontal" Connection Group Representations

There are four "horizontal" U(2) Connection groups labeled A, B, C, and D in Fig. 7-B.2. They each connect the Normal and Dark sectors of each of the four layers.

Their group representations contain a pair of corresponding fermions: from the Normal and Dark sectors. The corresponding fermions have identical quantum numbers except for the Darkness number which is 0 for the Normal fermion and 1 for the Dark fermion.

As a result there are 32 Connection group representations for each layer. In total there are 128 horizontal U(2) Connection group representations.

8.8 Quadplex Enhancement of the UST

The Quadplex enhancement of the UST in eq. 8.1, where fermion wave functions have four sets of coordinates, is a natural enhancement of the UST. It has the advantage of incorporating *instantaneous* Quantum Entanglement *explicitly* in the UST for fermion wave functions with tachyonic parts.

This enhancement can also be used to "improve" the UST dimension array map to symmetry groups. The seven Connection groups that we have used up to the present can be narrowed to the four horizontal Connection groups (A, B, C, D) in Fig. 7-B.2. The assignment of the E, F, and G U(2) groups to the set of Connection groups in Fig. 7.1 is modified to assignment to the z, u, and v coordinates with SL(2, **C**) group structure. See Fig. 8.5.

The four coordinate systems, y, z, u, and v, have four matching SL(2, C) groups[60] that are identical in the Normal and Dark sectors. The Dark sector uses the same coordinate systems since Dark matter/energy occupies well-defined positions in the universe.

Thus the Quadplex formulation is well-adapted to the UST.

---

[60] $SO^+(1, 3)$.

8.8.1 Quadplex Structuring and UST Layers

The Quadplex structuring in Fig. 8.5 raises the issue of the distribution of the coordinate systems' groups to four UST layers. The issue is resolved by considering the role of the layers. In the case of the fermions, the layers are distinguished by their great difference in masses. The first layer, the layer that we have observed, has smaller masses than the other layers of fermions with much higher masses. The mass spectrum is the source of the experimental differences.

The SL(2, **C**) groups for the coordinate sets have zero masses. Thus they may be considered on the same basis – irrespective of the layer in which they appear. The four sets of coordinate systems are associated with the layers as depicted in Fig. 8.5.

8.8.2 HyperCosmos Spaces of the Second Kind

The Second Kind HyperCosmos spaces do not have a Dark sector in our view. As a result there are no "horizontal" U(2) Connection groups labeled A, B, C, and D in Fig. 7-B.2. The U(2) "vertical" Connection groups E, F and G are replaced by SL(2, **C**) coordinate space groups for z, u, and v coordinate systems for use in the Lagrangian of eq. 8.1.

Thus there are no Connection groups for Second Kind HyperCosmos spaces. See Fig. 8.6.

8.9 The Intricate Interactions of the UST

There are 39 interactions in the UST, eight SU(4), eight ElectroWeak, eight Generation groups, eight Layer groups, and seven Connection groups. These groups result in a large number of fundamental representations involving all 256 fermions specified as a map from the Cosmos Theory UST dimension array. The number of fundamental representations for all interaction groups in the Dirac equation is 592 representations, which shows the depth of the dynamic Dirac equation analysis.

The interactions that they embody form an intricate pattern of interleaving vertical and horizontal parts when viewed from the perspective of the fermion dimension array. They reflect the principle that every fermion is connected to every other fermion through a series of interactions.

In the next chapter we consider the unification of interactions within our universe and the unification of all the interactions of all the spaces of Cosmos Theory.

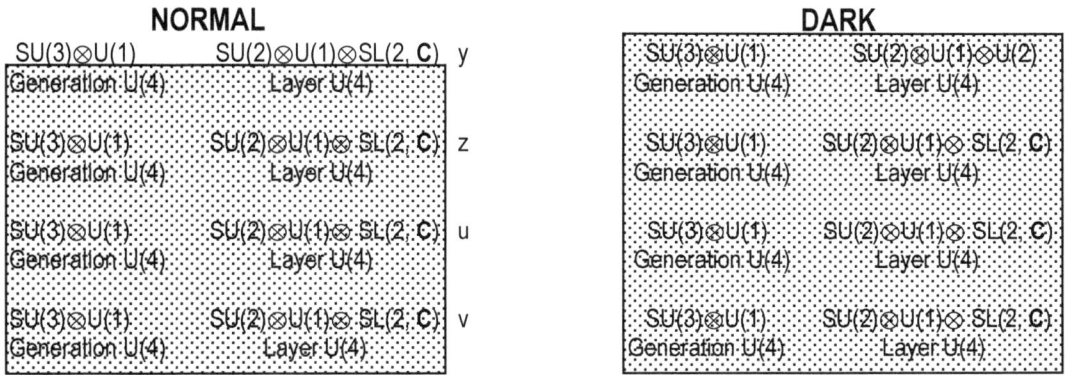

Figure 8.5. The transformed symmetries of the UST modified to the Quadplex formulation with SL(2, C) for the y, z, u, and v coordinate systems. The Dark sector uses the same coordinate systems since Dark matter/energy occupies well-defined positions in the universe. The darkened parts have not as yet been found experimentally. The one undarkened line is for experimentally known groups. The SL(2, C) representation has four coordinates.

Layer   Normal

4

3

2

1

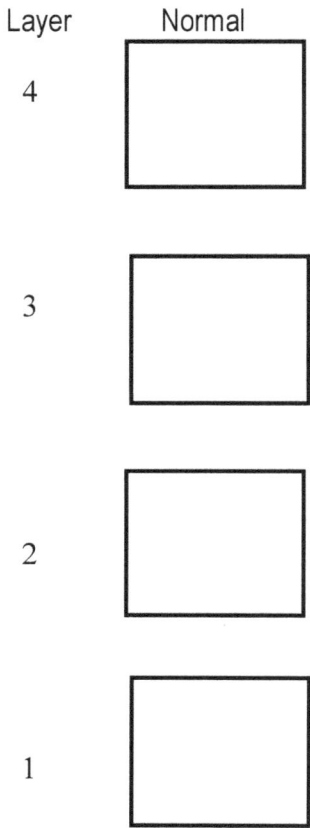

Figure 8.6. Second Kind HyperCosmos layers. There is no Dark sector.

9. Unification and Fundamental Reference Frames (FRF)

Unification of the interactions in quantum field theory particle models has been of interest for many years. In Cosmos Theory unification is easily achieved in our universe and in the universes of other Cosmos spaces.[61]

9.1 Unification in Our Universe

Unification in the quantum theory for our universe, the Unified SuperStandard Theory (UST), is implicit in the dimension array for the universe. The groups and interactions of the UST are automatically unified within the framework of the dimension array in the sense that all these groups may be viewed as subgroups of an overall group SU(256) for the dimension array *ab initio*.

A further explicit unification is possible with the introduction of Fundamental Reference Frames (FRF). Sets of interactions within their universes may be unified.

9.2 The FRF for Cosmos Spaces

Fundamental Reference Frames (FRF) provide a direct means of unifying interaction groups in such a way as to support rotations among the interactions. The general approach is to map the dimension array of a space to a vector. The rotations of the vector induce rotations of the dimension array and, consequently, rotations between the groups associated with the dimension array.

The ability to rotate vectors enables vectors to be transformed to having only one non-zero coordinate. Effectively this maps the set of groups associated with a dimension array to one trivial group. The process is comparable to transforming a momentum vector to a rest frame where there is only one non-zero component – the energy.

9.3 The Fundamental Reference Frame (FRF) as a "Rest Frame" for Reference Frames

A common feature of calculations in Physics is to choose a reference frame which facilitates computation. Common choices are the rest frame and the center of mass frame. In this chapter we find a Fundamental Reference Frame (FRF) (actually an infinite set of such frames) that plays a similar role for coordinate reference frames. We describe how to determine a Fundamental Reference Frame for each of the ten spaces (universes) of the HyperCosmos.

We suggest that a set of transformations exist that map from an FRF to a "normal" static reference frame. We are motivated in this endeavor by the observation made in previous books that the fermion particles of a space (universe) appear to be expressible as sets of replicated fermions. The replication of fermions is generated by transformations of an FRF subset of fermions to a set of replicates in the full FRF set.

---

[61] Much of this chapter is contained in Blaha (2023a).

The discussion begins with a view of our universe. In the r = 4 space[62] of our universe the 16 creation/annihilation b's and d's operators are

$$B = (b_{1\uparrow}, b_{2\uparrow}, b^\dagger_{1\uparrow}, b^\dagger_{2\uparrow},\ b_{1\downarrow}, b_{2\downarrow}, b^\dagger_{1\downarrow}, b^\dagger_{2\downarrow},\ d_{1\uparrow}, d_{2\uparrow}, d^\dagger_{1\uparrow}, d^\dagger_{2\uparrow},\ d_{1\downarrow}, d_{2\downarrow}, d^\dagger_{1\downarrow}, d^\dagger_{2\downarrow}) \quad (9.1)$$

in the PseudoQuantum formulation for the lowest spin fermion. A General Relativistic transformation T transforms B to a linear combination of b's and d's denoted B' as shown in our earlier work:

$$B' = TB \quad (9.2)$$

The transformation T has the form of a 16 by 16 matrix.

We now introduce an Internal Symmetry part with the goal of a complete unified theory. We add an index σ to each of the b's and d's where σ ranges from 1 through

$$2^{r/2+2} = 2^{2n-2} = 16 \quad (9.3)$$

where the space-time dimension is r = 4 and the Cayley number n = 3 for the N = 7, r = 4 space of our universe. The b's and d's number 256.

$$(b_{1\uparrow\sigma}, b_{2\uparrow\sigma}, b^\dagger_{1\uparrow\sigma}, b^\dagger_{2\uparrow\sigma},\ b_{1\downarrow\sigma}, b_{2\downarrow\sigma}, b^\dagger_{1\downarrow\sigma}, b^\dagger_{2\downarrow\sigma},\ d_{1\uparrow\sigma}, d_{2\uparrow\sigma}, d^\dagger_{1\uparrow\sigma}, d^\dagger_{2\uparrow\sigma},\ d_{1\downarrow\sigma}, d_{2\downarrow\sigma}, d^\dagger_{1\downarrow\sigma}, d^\dagger_{2\downarrow\sigma}) \quad (9.4)$$

The choice of the number of σ components, 16 in this case, is motivated by the goal of symmetry between the number of spin determined b and d components and the number of Internal Symmetry determined components—16 in both cases in this space. .

We now note the number of b's and d's are equal to the number of elements d_{dN} in the r = 4 dimension array.

$$d_{dN} = 2^{r/2+2} \cdot 2^{2n-2} = 2^{r+4} = 2^{4n-4} = 256 \quad (9.5)$$

The enhanced transformation of the b's and d's B_E has the form:

$$B_E' = T_E B_E \quad (9.6)$$

where T_E is a 256×256 array.

T_E is a General Relativistic – Internal Symmetry transformation.

9.3.1 HyperUnification Space

We now consider a related space of our r = 4 space. This space has transformations of the form of eq. 9.6 where the transformations are purely General Relativistic. Our goal for this space, which we call the HyperUnification space for our

---

[62] The variable r is the number of dimensions of a space or universe.

universe, is to unify space-time and internal symmetries. We wish it to have the form of HyperCosmos spaces.

In Blaha (2022e) and (2022g) we presented a unification for our universe based on a map to a higher space-time dimension r' HyperCosmos space called HyperUnification space.[63] This HyperUnification space has a General Relativity that generates transformations that map to forms of the dimension array $d_{dN'}$ where N' is the Blaha number of the HyperUnification space. The HyperUnification space and the "lower" space have space-time dimensions related by:

$$d_{dN} = 2^{22-2N} = 2^{r+4} \tag{9.7}$$

$$r' = 2r + 4 \tag{9.8}$$

based on the relation of the r dimension transformation and the r' dimension transformation. *Eq. 9.8 is the key for determining the HyperUnification space wherein the r dimension array becomes a vector in the r' space. In the HyperUnification space the rotations of the vector are purely GR transformations.*

General Justification of r' = 2r + 4

To obtain the dimension array for the r coordinate space-time we can use the r' coordinates space-time of the HyperUnification space. The r' space-time transformation has a $2^{r'/2+2} \times 2^{r'/2+2} = 2^{r+4} \times 2^{r+4}$ transformation array that generates a $2^{r'/2+2} = 2^{r+4}$ coordinates vector from its FRF. This vector maps to the r space-time dimension array d_{dN}.[64] The HyperUnification transformations are General Relativistic transformations of the HyperUnification space of dimension r'. *These transformations embody both the General Relativistic space-time and Internal Symmetries of the lower HyperCosmos space.*

The $2^{r'/2+2} = 2^{r+4}$ vector elements of the HyperUnification space become the elements of the d_{dN} dimension array of the r dimension space.

The defining feature of a HyperUnification space with space-time dimension r' of a space with a space-time dimension r space is that the number of components in the dimension r' GR transformation vector equals the number of components d_{dN} in the dimension array of the space of space-time dimension r.

For example, in our four dimension universe the HyperUnification space has 12 space-time coordinates. Its dimension array has 256×256 = 65,536 elements.[65] It has a 256-vector that is the d_{dN} = 256 elements dimension array[66] of our universe with 4

---

[63] See Blaha (2022d) and (2022g) for more detail. The higher dimension space may be one of the ten HyperCosmos spaces or it may be another HyperCosmos-like space. The unification spaces have space-time dimensions ranging up to 40 dimensions. See Fig. 9.1.

[64] The array is divided into a square d_{dN} array where Blaha number N = 9 – r/2.

[65] The value of the elements is not relevant for our discussions. The fact that they are non-zero in general is the key point.

[66] Again we note the value of the elements is not relevant. They play a symbolic role in being mapped to fermions and mapped to symmetry group dimensions, and so on.

space-time dimensions and 252 internal symmetry dimensions. The HyperUnification space-time dimension procedure yields the r space-time dimension array. *The key role of the HyperUnification space is Unification.*

Note:

     We have used the space corresponding to $r^{\centerdot}$ space-time dimensions to create the r space-time dimension array. The dimension array in each space-time specifies the number of fermion creation and annihilation operators in that space-time.

     The number of elements in the dimension array of the HyperUnification space of space-time dimension $r^{\centerdot}$ is:

$$d_{dN}' = 2^{r' + 4} \tag{9.9}$$

In terms of r' we find its Blaha number N' is

$$N' = \tfrac{1}{2}(18 - r^{\centerdot}) = 7 - r \tag{9.10}$$

and

$$N = 9 - r/2 \tag{9.11}$$

Note: negative values of N' are allowed for spaces where $r > 7$ in eq. 9.10.

THE HYPERCOSMOS SPACES SPECTRUM

Blaha Space Number **N**	Cayley-Dickson Number **n**	Cayley Number **d_c**	Dimension Array column length **d_{cd}**	Dimension Array Size **d_{dN}**	Space-time-Dimension **r**	Space-time r' Source of d_{dN} **r'**	HyperUnification Space Array **$d_{dN}{}^{t}$**
0	10	1024	2048	2048^2	18	40	2^{44}
1	9	512	1024	1024^2	16	36	2^{40}
2	8	256	512	512^2	14	32	2^{36}
3	7	128	256	256^2	12	28	2^{32}
4	6	64	128	128^2	10	24	2^{28}
5	5	32	64	64^2	8	20	2^{24}
6	4	16	32	32^2	6	16	2^{20}
7	3	8	16	16^2	4	12	2^{16}
8	2	4	8	8^2	2	8	2^{12}
9	1	2	4	4^2	0	4	2^{8}

Figure 9.1. The HyperCosmos spaces spectrum related to a HyperUnification space space-time dimension r' and its dimension array. Note the spaces with r' > 18 are outside the HyperCosmos set of 10 spaces. However they are made to have the same form as the 10 HyperCosmos spaces.

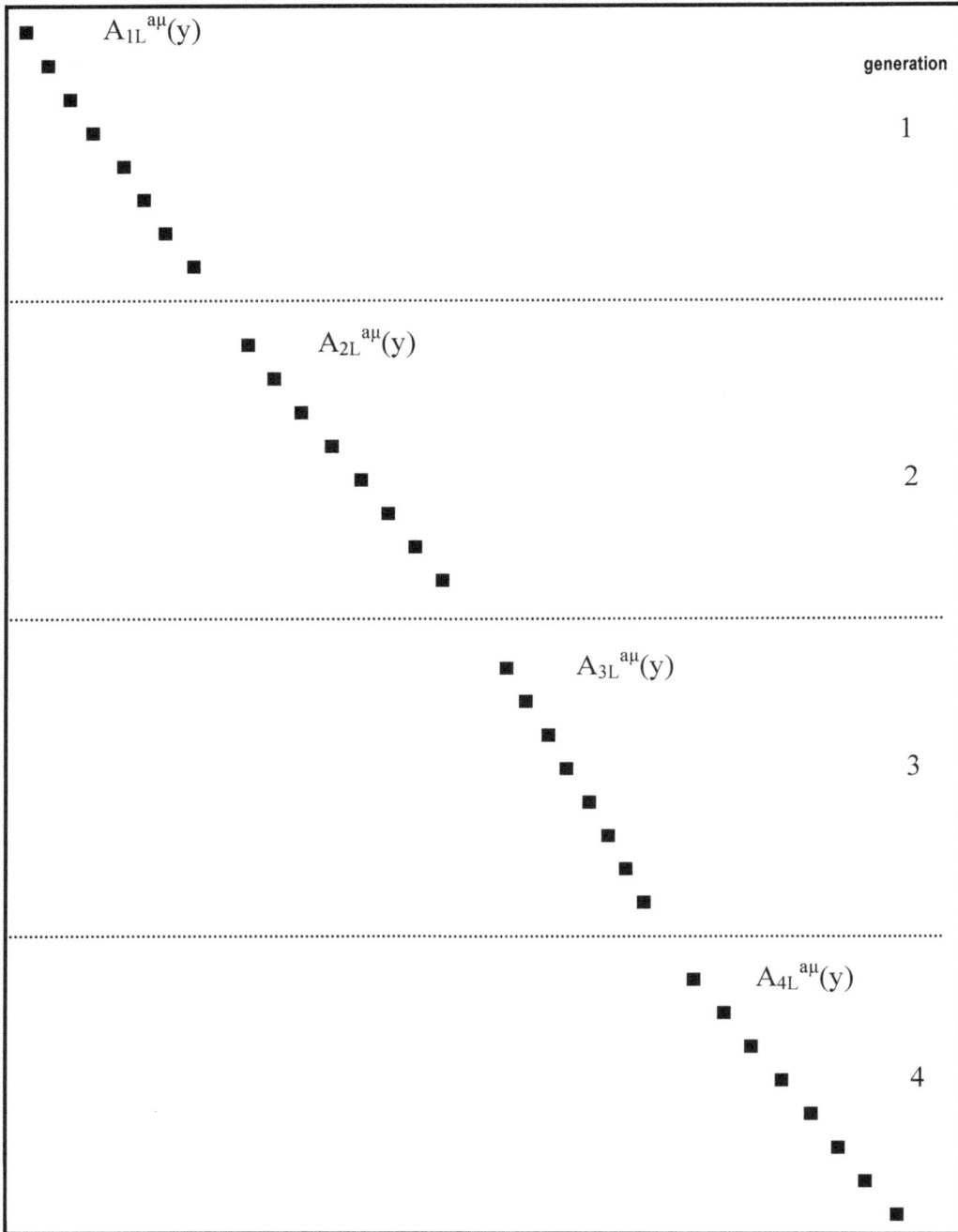

Figure 8.4. Eight species Layer group representations for each generation. We specify each four dimension Layer group representation with the ■ symbol. The representations in the figure are in block diagonal form. The number k of the Layer group $A_{kL}^{a\mu}(y)$ is chosen to be the number of the generation of its representations.

Contents of Fundamental Reference Frames

N = 9 - r/2	Number of Physical Elements	Total Number of Elements Generated
0	2^{11}	2^{22}
1	2^{10}	2^{20}
2	2^{9}	2^{18}
3	2^{8}	2^{16}
4	2^{7}	2^{14}
5	2^{6}	2^{12}
6	2^{5}	2^{10}
7	2^{4}	2^{8}
8	2^{3}	2^{6}
9	2^{2}	2^{4}

Figure 9.2. The contents of the ten types of Fundamental Reference Frames for the ten spaces. The elements are fermions or dimensions or symmetry fundamental representation dimensions as the case may be.

Contents of a Fundamental Reference Frame (FRF) for Fermions

N	Number of Physical Elements	Fermion Content	
0	2^{11}		
1	2^{10}		
2	2^{9}		
3	2^{8}	•	
4	2^{7}	•	
5	2^{6}	•	
6	2^{5}	**NORMAL** e q-up ν q-down e' q-up' ν' q-down' **DARK** e q-up ν q-down e" q-up" ν" q-down"	**Total: 32 fermions**
7	2^{4}	Normal: e q-up ν q-down Dark: e q-up ν q-down	**16 fermions**
8	2^{3}	e q-up **AND** ν q-down	**8 fermions**
9	2^{2}	e q-up **OR** ν q-down	**4 fermions**

Figure 9.3. Possible initial FRF contents of five of the ten types of Fundamental Reference Frames for fermions. The primes distinguish different fermions.

The initial FRF map to symmetry group contents in the Fundamental Reference Frame are listed for $N > 5$ below. Note the units of symmetry irreducible representations dimensions are four real-valued dimensions. Irreducible symmetry group units also have four dimensions. The number of units repeatedly doubles as we ascend up to $N = 0$.

Contents of a Fundamental Reference Frame for Symmetry Group Dimensions

N	Number of Physical Elements	Fermion Content	
0	2^{11}		
1	2^{10}		
2	2^{9}		
3	2^{8}	$\cdot$	
4	2^{7}	$\cdot$	
5	2^{6}	$\cdot$	
6	2^{5}	NORMAL U(4)⊗U(4)	Total: 32 real dimensions
		DARK U(4)⊗U(4)	
7	2^{4}	NORMAL U(4) ⊗ DARK U(4)	16 real dimensions
8	2^{3}	U(4)	8 real dimensions
9	2^{2}	U(2)	4 real dimensions

Figure 9.4. Possible map of initial FRF contents to five of the ten types of Fundamental Reference Frames for internal symmetry dimensions. The groups may undergo a further breakdown after being mapped

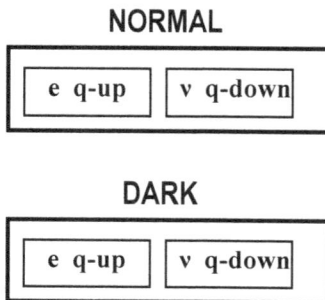

NORMAL

e q-up	v q-down

DARK

e q-up	v q-down

Figure 9.5. The 16 fermions that constitute the initial set that is then replicated 16 times to obtain the 256 fermion spectrum for N = 7 space (our universe).

The separation of the 16 fermions into two pairs of four fermions in Fig. 9.5 reflects the general separation of fermions and internal symmetries into 4, 8, 16, 32, and 64 parts due to symmetry splitting as we discussed in Blaha (2021b).

The set of symmetry groups in the UST for our universe also exhibits the same sort of replications seen for fundamental fermions. In the FRF there are 16 dimensions for U(4)⊗U(4) according to Fig. 9.4 for N = 7.

After a transformation, the UST symmetry groups have 256 real-valued dimensions for a product of the fundamental representations of their component groups. The basic symmetry group set for an N = 7 space is a U(4)⊗U(4) group. After a 16-fold replication the U(4) factors undergo transformations/breakdowns to

U(4) or SU(3)⊗U(1) or SU(2)⊗U(1)⊗SL(2, **C**)[68] or SU(2)⊗U(1)⊗U(2)

as seen below.

9.5 Replicates Generated by Transformations in HyperCosmos Spaces

The replication of the $2^{r/2+2}$ non-zero elements of an FRF is the result of a transformation in the HyperUnification as shown in eq. 9.6. Thus there is a replication of a replication in the FRF. This is visualized in Fig. 9.6.

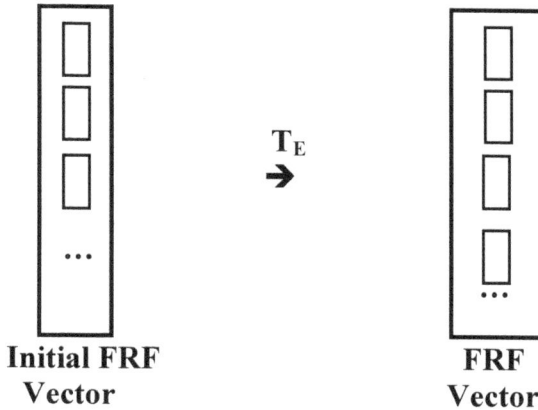

T_E
→

Initial FRF
Vector

FRF
Vector

Figure 9.6. Visualisation of the T_E transformation replication pattern in the HyperUnification space and its FRF. The FRF has $2^{r/2-2}$ replicates of the basic 16 dimension set.

The FRF of our N = 7 space has a $2^0 = 1$ copy of the basic set in the FRF vector. The purely General Relativistic HyperUnification transformation generates 16 replicates or 256 non-zero dimensions. These replicates may be visualized in Figs. 9.7 – 9.10 for our universe's UST.

---

[68] We use SL(2, C) to represent $SO^+(1, 3)$.

Number of Columns = 4   **NORMAL**   4   4   **DARK**   4

	NORMAL		DARK	
Layer 1 4	e 3 up-quarks	ν 3 down-quarks	e 3 up-quarks	ν 3 down-quarks
Layer 2 4	e 3 up-quarks	ν 3 down-quarks	e 3 up-quarks	ν 3 down-quarks
Layer 3 4	e 3 up-quarks	ν 3 down-quarks	e 3 up-quarks	ν 3 down-quarks
Layer 4 4	e 3 up-quarks	ν 3 down-quarks	e 3 up-quarks	ν 3 down-quarks

Figure 9.7. The UST 16×16 fermion spectrum of our universe tentatively arranged as SU(4)-plets that correspond directly with SU(4) (or SU(3)$\otimes$U(1)) fermions. There are four layers. Each set of 4 fermions has 4 generations matching the number of rows in each layer. This Periodic Table is broken into Normal and Dark sectors.

The Fermion Periodic Table (N = 7)

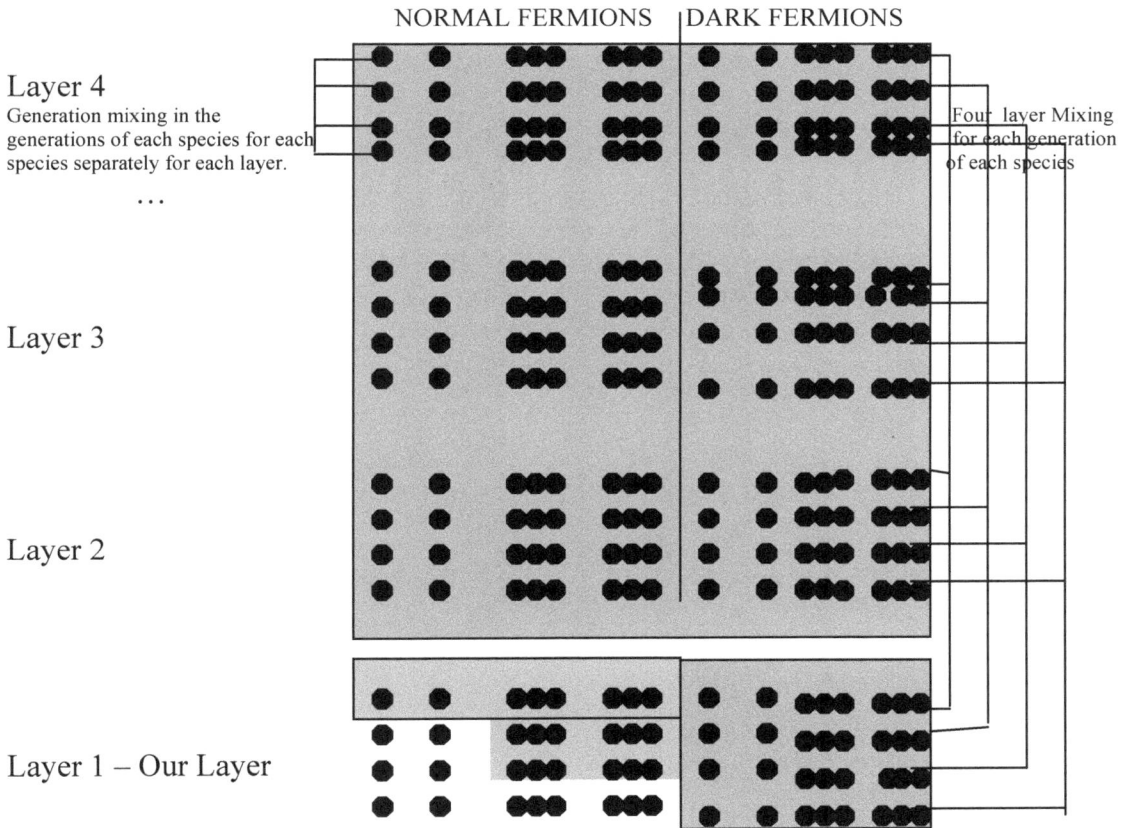

NORMAL FERMIONS   DARK FERMIONS

Layer 4
Generation mixing in the
generations of each species for each
species separately for each layer.

Four layer Mixing
for each generation
of each species

. . .

Layer 3

Layer 2

Layer 1 – Our Layer

Figure 9.8. UST Fermion particle spectrum and partial examples of the pattern of mass mixing of the Generation groups and of the Layer groups. Unshaded parts are the known fermions, The lines on the left side (only shown for one layer) display the Generation Group mixing within each layer. The Generation mixing occurs within each layer using a separate Generation group for each layer. The lines on the right side show Layer group mixing (for Dark matter) with the mixing among all four layers for each of the four generations individually. There are four Layer groups for Normal matter and four Layer groups for Dark matter. There are 256 fundamental fermions. The UST have the same fermion spectrum.

Initially the N = 7 UST full set of groups has the form of 16 $U(4) \otimes U(4)$ symmetry groups (Fig. 9.3.) Then they are transformed by symmetry breaking to the groups in Fig. 9.7.

	NORMAL	DARK
Layer 1	U(4)⊗U(4)	U(4)⊗U(4)
	U(4)⊗U(4)	U(4)⊗U(4)
Layer 2	U(4)⊗U(4)	U(4)⊗U(4)
	U(4)⊗U(4)	U(4)⊗U(4)
Layer 3	U(4)⊗U(4)	U(4)⊗U(4)
	U(4)⊗U(4)	U(4)⊗U(4)
Layer 4	U(4)⊗U(4)	U(4)⊗U(4)
	U(4)⊗U(4)	U(4)⊗U(4)

Figure 9.9. The "initial" distribution of sets of $N = 7$ symmetry groups. Each set is distinct and supports interactions only for the corresponding set of fermions (separately for Normal and Dark fermions). *Thus each set of 16 fermion generations has its own quantum numbers and interactions.* Each U(4)⊗U(4) set has a 16 real-valued dimension representation, which is importance when we consider Fundamental Reference Frames.

NORMAL		DARK	
SU(3)⊗U(1)	SU(2)⊗U(1)⊗SL(2, C)	SU(3)⊗U(1)	SU(2)⊗U(1)⊗U(2)
Generation U(4)	Layer U(4)	Generation U(4)	Layer U(4)
SU(3)⊗U(1)	SU(2)⊗U(1)⊗U(2)	SU(3)⊗U(1)	SU(2)⊗U(1)⊗U(2)
Generation U(4)	Layer U(4)	Generation U(4)	Layer U(4)
SU(3)⊗U(1)	SU(2)⊗U(1)⊗U(2)	SU(3)⊗U(1)	SU(2)⊗U(1)⊗U(2)
Generation U(4)	Layer U(4)	Generation U(4)	Layer U(4)
SU(3)⊗U(1)	SU(2)⊗U(1)⊗U(2)	SU(3)⊗U(1)	SU(2)⊗U(1)⊗U(2)
Generation U(4)	Layer U(4)	Generation U(4)	Layer U(4)

Figure 9.10. The transformed/broken sets of symmetries in UST and in $N = 7$, $r = 4$ HyperCosmos space. Note each element has a 16 real dimension representation. This depiction is also evident in the UST. The SL(2, **C**) representation has four coordinates.[69]

9.6 Portrait of a HyperCosmos Space and its Associated HyperUnification Space

Fig. 9.11 shows the relation between a HyperCosmos space and its related HyperUnification space and their FRFs.

---

[69] The Lorentz Group $SO^+(1, 3)$ is often specified with an SL(2, **C**) representation.

HyperUnification Space
N'
r'

N' **Fundamental
Reference Frame**

T_E

N **Fundamental
Reference Frame**

N HyperCosmos Space
r

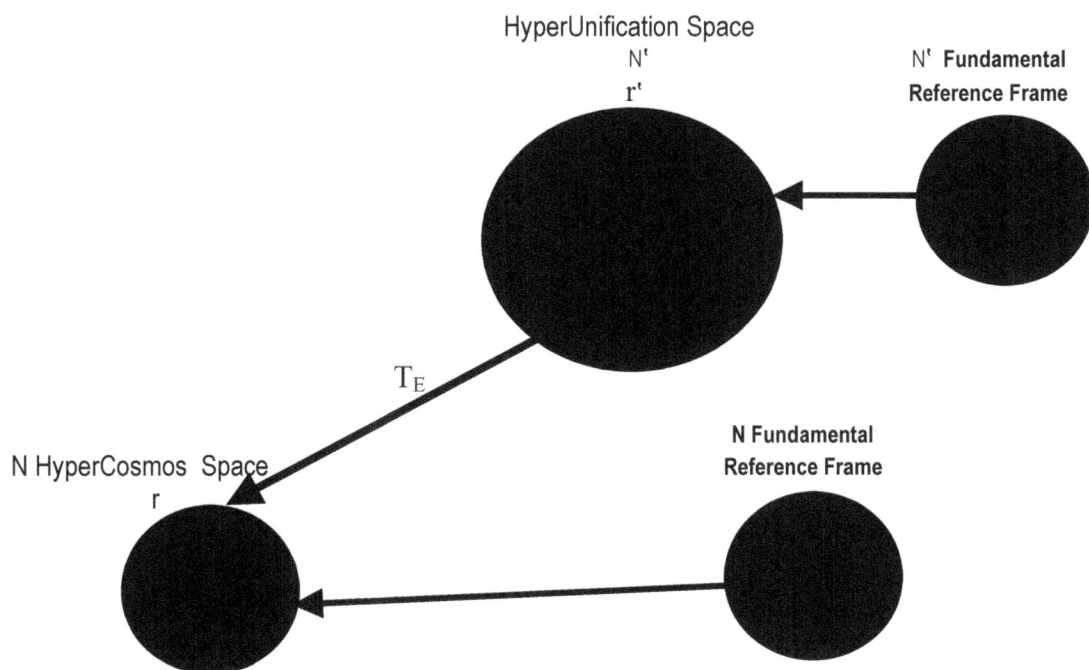

Figure 9.11. Diagram of the relation between a HyperCosmos space and its associated HyperUnification space with Fundamental Reference Frames indicated. The HyperUnification space sets the dimension array d_{dN} for the HyperCosmos space using purely HyperUnification General Relativity. Fundamental Reference Frames are like the rest frames of particles. They reduce the number of elements to a core set of elements. The elements are dimensions or fermions or gauge vector symmetry fundamental representation dimensions.

9.7 Reduction of an FRF's Contents to One Non-Zero Dimension in a HyperUnification Space

A HyperUnification Space for $r \geq 4$ has an FRF with $2^{r/2-2}$ replicates of the 16 non-zero elements in our universe's FRF. The 16 elements corresponded to the 16 b$\wedge$'s and d's of a spin ½ fermion in our universe. The remainder of the FRF vector is $2^{r+4} - 2^{r/2+2}$ zeroes.

Transformations of the HyperUnification space also support the generation of the *entire* set of HyperCosmos dimensions from one dimension in its FRF vector in the HyperUnification Space.

We demonstrate this possibility with a simple example. Assuming the FRF has one non-zero dimension (element) in row 1 of the FRF vector with all other components zero we define a vector with d_{dN} components as

$$V = (a, 0, 0, \ldots) \qquad (9.12)$$

and a prototype General Relativistic transformation

$$[T]_{ij} = b_i \ \delta_{i1} + (1 - \delta_{1i})(1 - \delta_{j1})c_{ij} \qquad (9.13)$$

where b_i and the c_{ij} are numeric. Transformation T has entries b_i in column 1 and a matrix c_{ij} in the rows and columns for i, j = 2, 3, ... A full set of d_{dN} array dimensions are generated from one FRF dimension (element). The transformation

$$V' = TV \qquad (9.14)$$

generates the vector V' with no zero-valued components:

$$V' = (a_1, a_2, a_3, \ldots , a_{d_{dN}}) \qquad (9.15)$$

This vector represents the d_{dN} dimension array when it is restructured into a square array. The representation is not numeric since dimensions have no numeric value. Each non-zero value in the dimension array signifies it is an *active* dimension that is used in determining a fundamental fermion spectrum or in determining the dimensions of an irreducible symmetry group representation.

Thus we can generate the complete set of dimensions in each HyperCosmos dimension array from one dimension in its FRF using a HyperUnification space General Relativistic transformation.

9.8 FRFs and Replicates

The $2^{r/2+2}$ non-zero elements in an FRF are generated from one element by a HyperUnification transformation. These elements consist of $2^{r/2-2}$ replicates of the 16 elements of our universe's FRF. Each replicate has different groups from the other replicates.

A further HyperUnification transformation maps the FRF contents to the d_{dN} elements that become a HyperCosmos dimension array when expressed as a square array.[70] This second transformation creates $2^{r/2+2}$ replicates of each of the FRF's $2^{r/2+2}$ elements (the $2^{r/2-2}$ replicates of the 16 elements in the FRF).

Thus the dimension array that emerges consists of 2^{r+4} elements distributed in 2^r replicates of the 16 elements of our universe's FRF. Each of these replicates maps to different sets of fundamental fermions or to different sets of irreducible representations of symmetry groups.

We see these replicates in the UST figures for our universe! Chapter 5 analyzes the 16 elements of our universe's FRF and shows that they reflect separations into sublight and superluminal parts based on the two time dimensions in our 12 space-time dimension HyperUnification space, which is equivalent to F-Theory space. See Figs. 9.7 and 9.10.

---

[70] The pair of transformations may be combined into one transformation from one element to the d_{dN} element vector that becomes the dimension array of the HyperCosmos space.

10. The Full HyperUnification Space

We now consider the combined unification[71] of all or some of the HyperCosmos spaces in a *Full HyperUnification Space* that will turn out to be a 42 space-time dimension space. Each space has a dimension array with d_{dN} elements. This array is reexpressed as a vector v_N with d_{dN} components residing in the space's unification space with space=time dimension r' given by eq. 9.8.

Now consider the sum of all ten HyperUnification spaces vectors:

$$v_S = \sum_{N=0}^{9} v_N \qquad (10.1)$$

$$= 4^2 + 8^2 + 16^2 + 32^2 + 64^2 + 128^2 + 256^2 + 512^2 + 1024^2 + 2048^2$$

with[72]

$$v_N = d_{dN} \qquad (10.2)$$

These elements form a v_S-vector in the Full HyperUnification space. See Figs. 10.1 and 10.2. The sum of the ten vectors of the ten HyperUnification spaces vectors is a vector d_{cS} of the Full HyperUnification Space:

$$d_{cS} = v_S = 5,592,400 \qquad (10.3)$$

There is a corresponding set of square dimension array blocks whose total is

$$d_{dS} = 4^4 + 8^4 + 16^4 + 32^4 + 64^4 + 128^4 + 256^{4\wedge} + 512^4 + 1024^4 + 2048^4 = 1.8765 \times 10^{13} \qquad (10.4)$$

The Full HyperUnification Space plays the role of unification for all ten HyperCosmos spaces. Each of the ten HyperUnification space vectors of size d_{dN} in v_S individually has d_{dN} components. Correspondingly, each subvector has a $d_{dN} \times d_{dN}$ square array along the diagonal of a combined transformation which we call a *HyperUnification Transformation*. (Fig. 10.2)

Using eq. 9.8 we find the dimension array block defined by the d_{cS} vector is

$$d_H = 5,592,400^2 = 3.12749 \times 10^{13} \qquad (10.5)$$

If we treat this square array d_H as corresponding to a HyperCosmos space then we can calculate the corresponding space-time dimension

---

[71] Much of this chapter is contained in Blaha (2023a).
[72] Note Blaha number N = 9 – r/2.

$$r_H = \log_2(d_H/16) \qquad (10.6)$$
$$= 2\log_2(5{,}592{,}400) - 4 = 40.83$$

using

$$d_{dH} = 2^{22-2N_H} = 2^{r_H+4} \qquad (10.7)$$

The dimension r_H is non-integral. Noting that space-time dimensions are even numbered in HyperCosmos spaces. we define the *Full HyperUnification Space* to have

$$r_H = 42 \qquad (10.8)$$

space-time dimensions. It is the smallest space-time dimension that has a dimension array block containing the d_H dimension array block as shown in Fig. 10.2. (Note: the dimension array block d_{drH} has the same size as the General Relativistic transformation between the FRF and a static space-time.)

10.1 General Relativistic Transformation

The General Relativistic transformations of the $r_H = 42$ Full HyperUnification Space includes the General Relativistic transformations of any or all, HyperUnification spaces along its "diagonal" where blocks correspond to individual unification space General Relativistic transformations. See Fig. 10.1. It also includes much more in the form of arbitrary, not necessarily block diagonal, transformations that mix various HyperCosmos unification spaces.

The 42 space-time dimension array has the vector column size

$$d_{crH} = 2^{42/2+2} = 2^{23} = 8{,}388{,}608 \qquad (10.9a)$$

with the number of array elements being

$$d_{drH} = 8{,}388{,}608^2 = 7.03687 \times 10^{13} \qquad (10.9b)$$

Since d_{crH} is larger than d_{cS} there is an apparent excess in the 42 space-time dimension Full HyperUnification Space which we consider below. (See Fig. 10.2.)

We can define a set of HyperUnification Space General Relativistic transformations that transform each of the individual HyperUnification sub-blocks separately. We can also define transformations that transform combinations of all 10 HyperUnification spaces including the entire set of HyperUnification spaces.

10.2 Full HyperUnification Space FRF

The FRF vector for the Full HyperUnification Space has d_{crH} components. Comparing it to the size of $d_{cS} = 5{,}592{,}400$-vector we find the General Relativistic transformations of the 10 HyperCosmos unification spaces are a subset of the 42 space-time dimension HyperUnification Space transformations. The column excess is

2,796,208 dimensions. The ratio of the larger box labeled A to the smaller box labeled B in Fig. 1.2 is exactly

$$d_{crH}/d_{cS} = 1.5 = 3/2 \tag{10.10}$$

See Figs. 10.1 – 10.3 for the relations of the various parameters. Thus the B box column size d_{cB} satisfies

$$d_{cB} = \tfrac{1}{2}\, d_{cS} \tag{10.11}$$

or

$$d_{cB} = \tfrac{1}{2} \sum_{N=0}^{9} v_N \tag{10.12}$$

Eq. 10.12 suggests that the excess of space within the 42 dimension space has a relevant interpretation. We explore it in chapter 11. It suggests that the excess—the B block— is a scaled down version of the A block. We call it the Full HyperUnification Space of the Second Kind. It corresponds to a new *HyperCosmos of the Second Kind* described in chapter 11.

   The fact that the 42 dimension space directly contains what will be seen to be **exactly** *two HyperCosmos's supports the choice of a 42 dimension space.*

   We will use HyperUnification Space transformations in the next chapter to reduce the set of all HyperCosmos spaces' dimensions to one universal, primordial dimension.

   We note that the ratio of the square dimension arrays labeled A and B in Fig. 10.2 is exactly

$$d_{dNH}/d_{dS} = 7.03687 * 10^{13}/ 1.8765E+13 \tag{10.13}$$
$$= 3.75 = 15/4$$

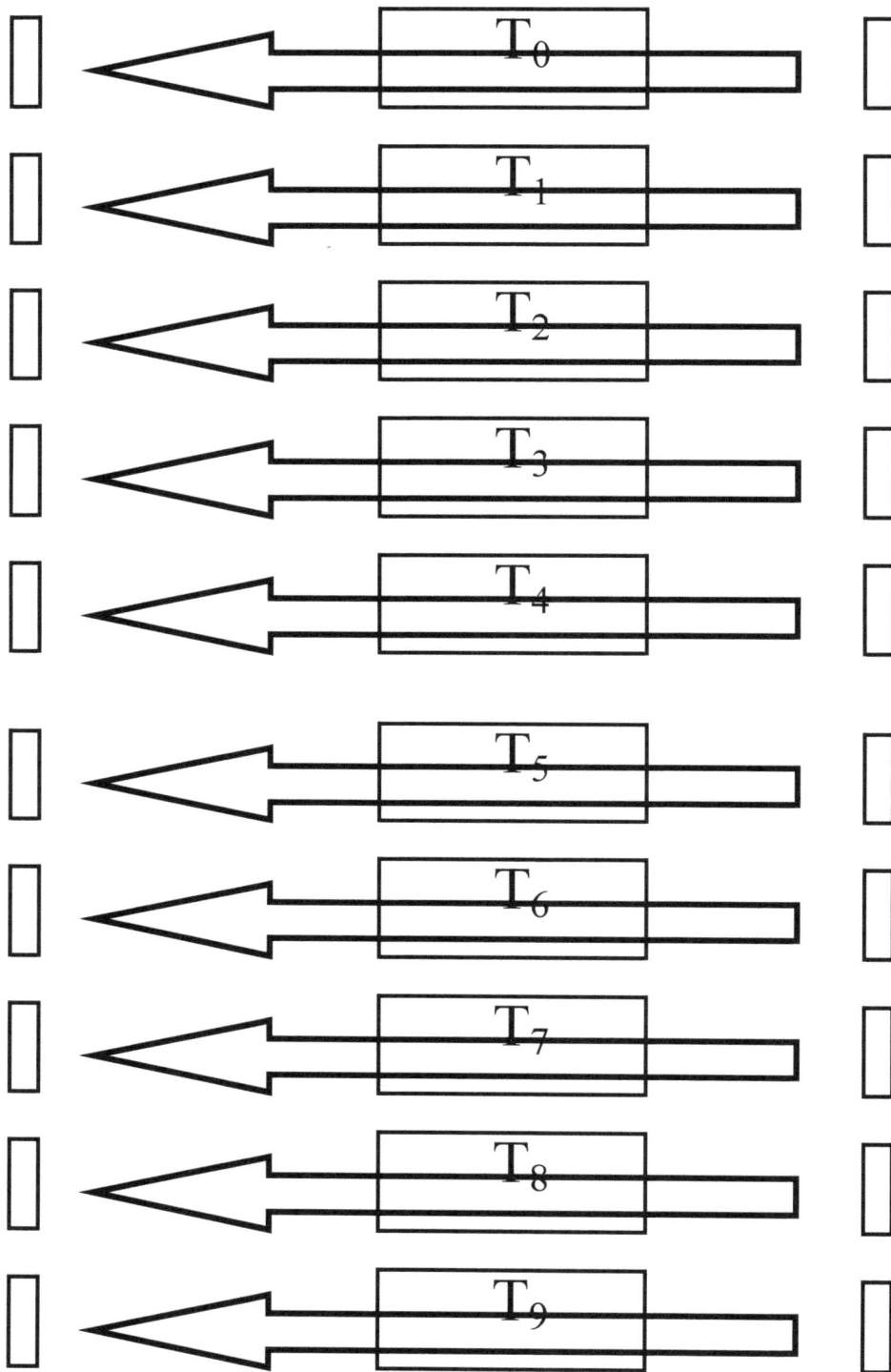

Figure 10.1. HyperCosmos FRF transformations in the Full HyperUnification Space.

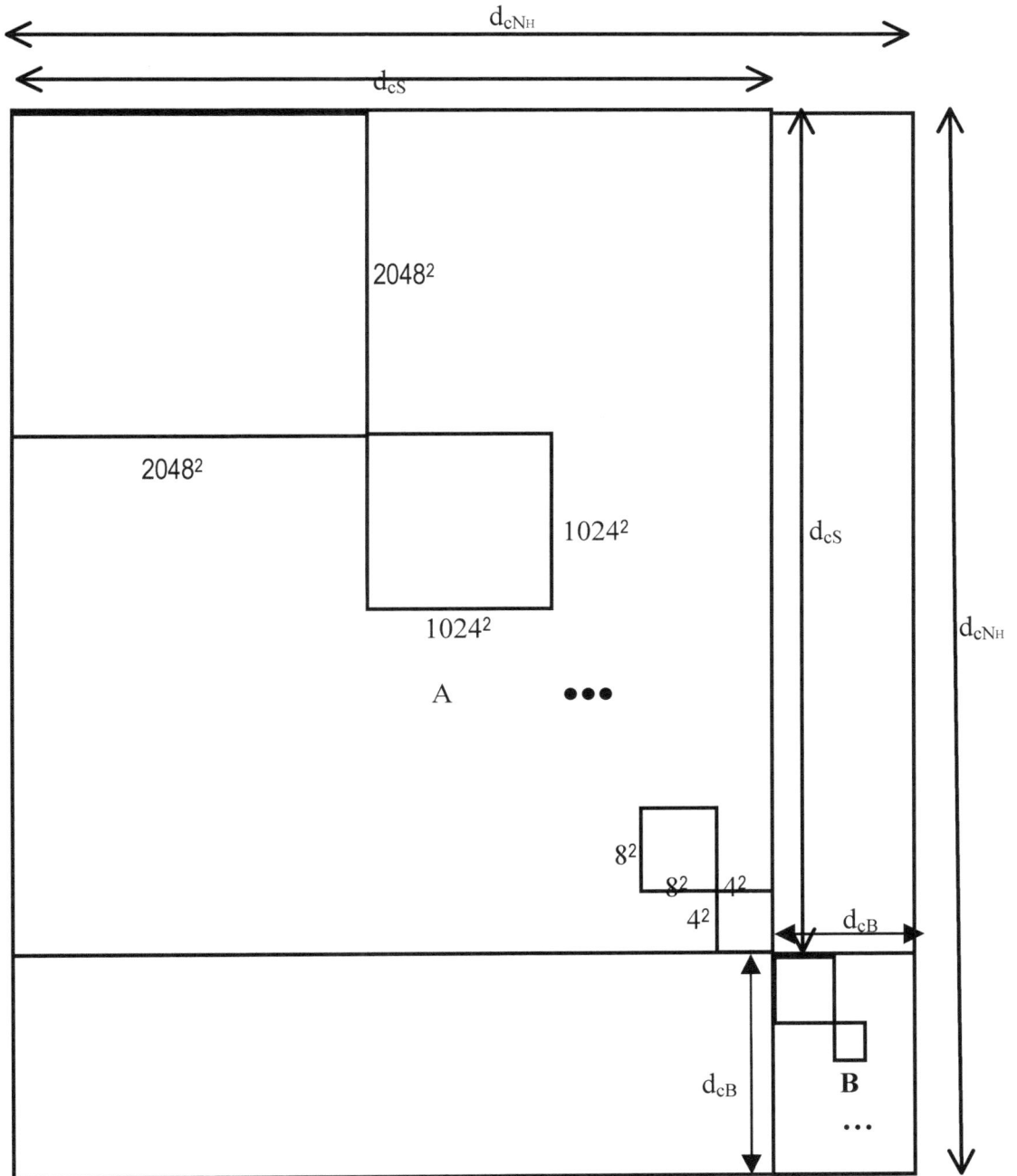

Figure 10.2. Form of a HyperUnification transformation. It is also the form of the dimension array of 42 dimension space-time. Block A is square. The figure is not drawn to scale.

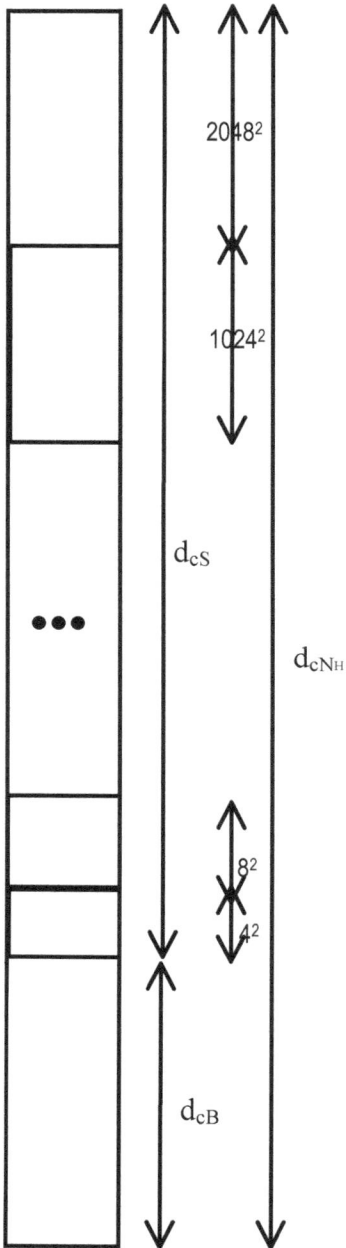

Figure 10.3. The r = 42 space-time HyperUnification vector form.

11. The HyperCosmos Spaces of the Second Kind

The HyperCosmos Spaces of the Second Kind emerges from the consideration of Full Unification in chapter 10. The form of d_{cR} for array B in chapter 10 suggests that the HyperCosmos of the Second Kind spectrum of spaces is similar to our original HyperCosmos with one basic change in the numerics:

$$d_{cr} = \tfrac{1}{2} \sum_{N=0}^{9} v_N = \tfrac{1}{2} d_{cS} = d_{cB} \qquad (11.1)$$

by eqs. 10.1, 10.3 and 10.11. The vector v_N

$$v_N = d_{dN} = d_{dr} \qquad (11.2)$$

using Blaha number $N = 9 - r/2$ should be modified to

$$v_{N'} = \tfrac{1}{2} d_{dN} = d_{dN'} = d_{dr'} \qquad (11.3)$$

which suggests

$$d_{dr'} = d_{dN'} = \tfrac{1}{2}\, 2^{r/2+2} 2^{r/2+2} = 2^{r/2+1} 2^{r/2+2} = 2^{r+3} \qquad (11.4)$$

The Second Kind arrays are not square. They are rectangular.
There are several ways to obtain a $2^{r/2+1}$ factor:

1. Change the internal symmetry factor by a factor of two. This can be accomplished by eliminating the Dark sector: no Dark fermions and no Dark interactions. This would explain the existence of Dark matter without any interactions except gravity.

2. Remove the PseudoQuantum quantum field framework. This approach would remove the advantages of PseudoQuantum field theory.

3. Restrict General Relativistic (GR) transformations to b and $b^\dagger$ operators only. The d and $b^\dagger$ operators would not undergo GR transformations. These operators are half of the vector operators in GR transformations.

It appears that choice 1 is the best since it conforms to the known absence of Dark matter features and best preserves a HyperCosmos formalism.
Thus a HyperCosmos of the Second Kind is possible. Whether Nature chooses it is an open question. We shall continue using the original HyperCosmos formalism.

The 42 space-time dimension space has a General Relativity that can mix the HyperCosmos spaces and the Second Kind HyperCosmos spaces. Thus there are gravitational interactions between them if the Second Kind set of spaces is instantiated with universes. We regard the HyperCosmos and the Second Kind HyperCosmos as siblings with the only interaction between them being gravitation at best.

The particles and symmetry groups of the Second Kind HyperCosmos spaces are not the same as the particles and symmetry groups in HyperCosmos spaces. The following figures show features of the $N = 7$ ($r = 4$) Second Kind HyperCosmos space and thus a Second Kind UST.

11.1 Two HyperCosmoses

We see that there are two HyperCosmoses: the original HyperCosmos and the HyperCosmos of the Second Kind. Either or both may be instantiated with universes. Our universe, at the moment, may be of either HyperCosmos type. The key undecided question is the existence of a Dark sector.

The form of the Full HyperUnification Space supports two HyperCosmoses exactly. It provides significant support for the choice of a 42 dimension space-time.

Whether our universe is in the original HyperCosmos or in the Second Kind HyperCosmos is not known at present. The choice depends on the existence of the Dark sector.

11.2 Types of Transformations

The 42 space-time dimension Full HyperUnification space General Relativistic transformations are:

1. Transformations on the FRF of each HyperCosmos HyperUnification space individually.

2. Transformations on the FRF of each Second Kind HyperCosmos HyperUnification space individually.

3. Combined transformations on the FRFs of HyperCosmos HyperUnification spaces.

4. Combined transformations on the FRFs of Second Kind HyperCosmos HyperUnification spaces.

5. Transformations on the complete set of Full HyperUnification Space FRFs of both HyperCosmos and Second Kind HyperCosmos HyperUnification spaces.

11.3 Second Kind HyperUnification Spaces

Each of the ten Second Kind HyperCosmos spaces has a corresponding HyperUnification space. We assume the set of space-time dimensions of the Second Kind HyperCosmos is the same as the HyperCosmos. For space-time dimension r there number 2^{r+3} by eq. 11.4. Consequently the corresponding Second Kind HyperUnification space has vectors with 2^{r+3} components as indicated in Figs. 6.2 and 7.1.

Since the relation of the space-time dimensions is

$$2^{r+3} = 2^{r'/2+2} \qquad (11.5)$$

for HyperCosmos-like spaces we find

$$r' = 2r + 2 \qquad (11.6)$$

where r' is the space-time dimension of the HyperUnification space.

HYPERCOSMOS OF THE SECOND KIND SPACES SPECTRUM

Blaha Space Number $N = O_s$	Cayley-Dickson Number n	Cayley Number d_c	Dimension Array size d_d	Space-time-Dimension r	CASe Group $su(2^{r/2}, 2^{r/2})$ CASe
0	10	1024	1024×2048	18	su(512,512)
1	9	512	512×1024	16	su(256,256)
2	8	256	256×512	14	su(128,128)
3	7	128	128×256	12	su(64,64)
4	6	64	64×128	10	su(32,32)
5	5	32	32×64	8	su(16,16)
6	4	16	16×32	6	su(8,8)
7	3	8	8×16	4	**su(4,4)**
8	2	4	4×8	2	su(2,2)
9	1	2	2×4	0	su(1,1)

Figure 11.1. The HyperCosmos of the Second Kind space spectrum. The space for our universe, is Blaha number N = 7, with Cayley-Dickson number 3 (which corresponds to octonions) is in bold type. Note changed d_d column relative to the HyperCosmos.

Number of Columns = 4   **NORMAL**   4   4 DARK 4

Layer 1 4	e 3 up-quarks	ν 3 down-quarks	e 3 up-quarks	ν 3 down-quarks
Layer 2 4	e 3 up-quarks	ν 3 down-quarks	e 3 up-quarks	ν 3 down-quarks
Layer 3 4	e 3 up-quarks	ν 3 down-quarks	e 3 up-quarks	ν 3 down-quarks
Layer 4 4	e 3 up-quarks	ν 3 down-quarks	e 3 up-quarks	ν 3 down-quarks

Figure 11.2. The HyperCosmos of the Second Kind 8 × 16 fermion spectrum tentatively arranged as SU(4)-plets that correspond directly with SU(4) (or SU(3)⊗U(1)) fermions. There are four layers. Each set of 4 fermions has 4 generations matching the number of rows in each layer. This Periodic Table is broken into Normal and Dark sectors. The absence of a Dark sector is indicated by the darkening.

	NORMAL	**DARK**
Layer 1	U(4)⊗U(4) U(4)⊗U(4)	U(4)⊗U(4) U(4)⊗U(4)
Layer 2	U(4)⊗U(4) U(4)⊗U(4)	U(4)⊗U(4) U(4)⊗U(4)
Layer 3	U(4)⊗U(4) U(4)⊗U(4)	U(4)⊗U(4) U(4)⊗U(4)
Layer 4	U(4)⊗U(4) U(4)⊗U(4)	U(4)⊗U(4) U(4)⊗U(4)

Figure 11.3. The Second Kind "initial" distribution of sets of Blaha number N = 7 symmetry groups. Each set is distinct and supports interactions only for the corresponding set of fermions (separately for Normal and Dark fermions). *Thus each set of 16 fermion generations has its own quantum numbers and interactions.* Each U(4)⊗U(4) set has a 16 real-valued dimension representation, which are of importance when we consider Fundamental Reference Frames. The absence of a Dark sector as indicated by the darkened part.

NORMAL

SU(3)⊗U(1)	SU(2)⊗U(1)⊗SL(2, C)
Generation U(4)	Layer U(4)
SU(3)⊗U(1)	SU(2)⊗U(1)⊗U(2)
Generation U(4)	Layer U(4)
SU(3)⊗U(1)	SU(2)⊗U(1)⊗U(2)
Generation U(4)	Layer U(4)
SU(3)⊗U(1)	SU(2)⊗U(1)⊗U(2)
Generation U(4)	Layer U(4)

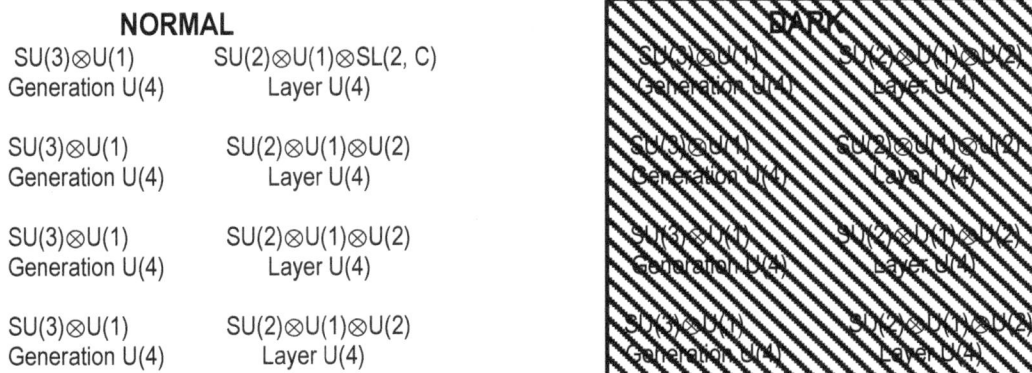

Figure 11.4. The transformed/broken sets of symmetries in the Blaha number N = 7, r = 4 Second Kind HyperCosmos space. Note each element has a 16 real dimension representation. This depiction is also evident in a Second Kind UST. The SL(2, C) representation has four coordinates.[73] The absence of a Dark sector as indicated by the darkened part.

---

[73] The Lorentz Group $SO^+(1, 3)$ is often specified with an SL(2, C) representation.

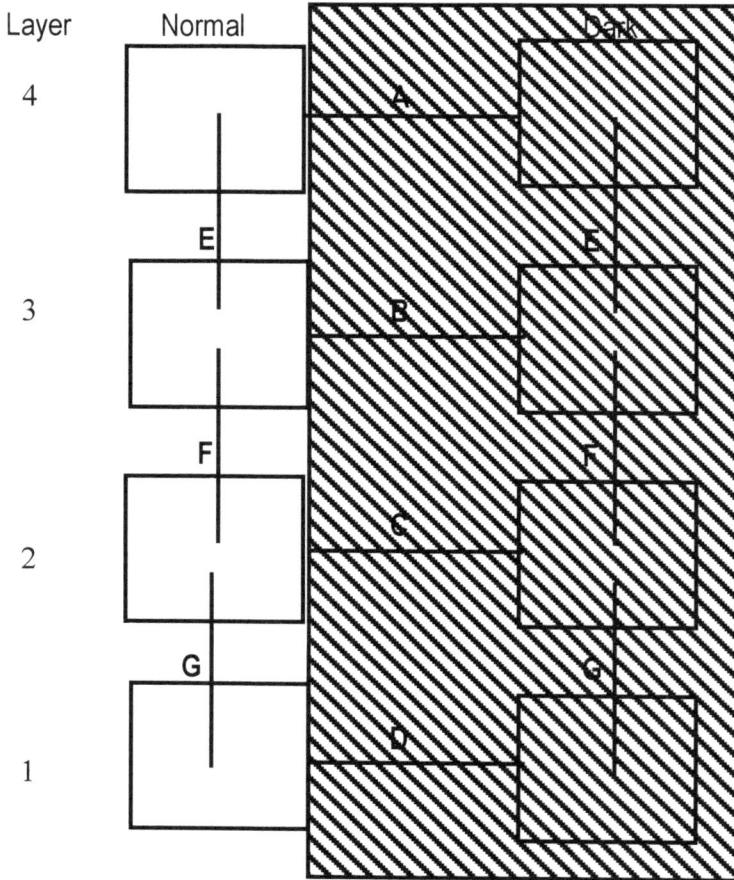

Figure 11.5. The three U(2) Connection groups[74] (shown as 3 lines) between the eight UST blocks in the Blaha number N = 7 Second Kind HyperCosmos. The Darkened part is not present in the Second Kind case. Connection groups are obtained by transfering 12 dimensions from the UST space-time to internal symmetries with the consequent reduction of the space-time from four octonion (complex quaternion) coordinates to four real coordinates. The Connection groups generate rotations and interactions between corresponding fermions and vector bosons of each pair of blocks. This figure is abstracted from the corresponding figure in Appendix 7-B.

---

[74] Connection groups are discussed in Appendix 9-B.

12. The UltraUnification (UU) Space[75] of the Full HyperUnification Space

The Full HyperUnification Space is a 42 space-time dimension space. It has a set of transformations of the form of Fig. 10.2. It also has a square dimension array of the same form as the transformation array: The number of rows (and columns) is d_{c42}. The 42 dimension array has a set of dimension arrays of HyperUnification spaces of the HyperCosmos and the Second Kind HyperCosmos corresponding to the blocks of Fig. 10.2.

We can specify a set of unified transformations of these blocks in the same manner as we originally did for HyperCosmos spaces when we defined HyperUnification spaces using

$$r' = 2r + 4 \tag{9.8}$$

with r = 42. The space-time dimension of its HyperUnification space, which we call the *UltraUnification (UU) Space*, is r' = 88. An 88 dimension General Relativistic transformation transforms all the contents of the Full HyperUnification Space. It therefore unifies all the space-time and Internal Symmetries within the set of HyperCosmos spaces and the Second Kind HyperCosmos spaces.

The UU space is needed to have an FRF with one non-zero dimension that can be transformed to the full set of dimensions in the 42 space-time dimension Full HyperUnification space and thence to the complete set of dimensions in the ten HyperCosmos spaces and in the ten Second Kind HyperCosmos spaces.

Note:

The defining feature of a HyperUnification space with space-time dimension r' for a space with space-time dimension r is that the r' HyperUnification dimension array vector solely undergoes strictly GR transformations and does not use internal symmetry transformation parts.

The 88 dimension UltraUnification space completes the Cosmos.[76] The Cosmos has four levels as shown in Fig. 12.1. The total number of Cosmos spaces is

!0 HyperCosmos spaces + their 10 HyperUnification spaces +   (12.1)
+ !0 Second Kind HyperCosmos spaces +

---

[75] Much of this chapter is contained in Blaha (2023a).
[76] The appearance of the dimensionless integers 42 and 88 in this Cosmos evokes the memory of the Eddington and Dirac (and others) conjectures on the appearance of 40 and 80 approximately in various combinations of universe parameters and physical constants. If some of their conjectures are valid they would be evidence for the author's Cosmos Theory. The dimensionless numbers 40 and 80 share the property of dimensionlessness with the dimensions of Cosmos Theory.

> +their 10 Second Kind HyperUnification spaces +
> + the 42 Dimension Full HyperUnification Space +
> + the 88 Dimension UU space
> = 42 spaces

We thus find the interesting equality of the Full HyperUnification space dimension 42 and the total number of Cosmos spaces. Whether this equality reflects a deeper underlying relation within the Cosmos' structure remains to be determined. Fig. 12.1 displays the four levels and the 42 spaces of the complete author's *Cosmos Theory*.

12.1 Relation Between the Four Levels

The four levels of the Cosmos (Fig. 12.1) are intimately related:

A. The fourth level consists of the ten HyperCosmos spaces augmented by ten Second Kind HyperCosmos spaces.

B. The third level consists of 20 HyperUnification spaces. They are the HyperUnification spaces of the HyperCosmos and of the Second Kind HyperCosmos. The fourth level spaces each have a second level HyperUnification space. General Relativistic transformations of the HyperUnification space generate the dimension array of each r dimension HyperCosmos and Second Kind HyperCosmos space from a vector in the space's FRF. The vector may have only one non-zero dimension or may have a set of $2^{r/2+2}$ non-zero dimensions (or some other number of non-zero dimensions).

C. The second level is a 42 space-time dimension space with the same overall form as a HyperCosmos space. In this space the 20 HyperUnification spaces of the third level appear "along the diagonal."

D. The first level is the 88 space-time dimension UltraUnification space of the third level 42 dimension space. General Relativistic transformations of the UltraUnification space generate vectors yielding the dimension array of the 42 dimension second level HyperCosmos space from a vector in the first level space's FRF. The vector may have only one non-zero dimension or a set of $2^{42+4} = 2^{46}$ non-zero dimensions (or some other number of non-zero dimensions).

A General Relativistic transformation of the 88 space-time dimension UltraUnification space cascades down to the fourth level HyperCosmos and Second HyperCosmos dimension arrays. Consequently it is possible to populate all HyperCosmos and Second Kind HyperCosmos dimension arrays with one non-zero dimension in the FRF vector of the 88 space-time dimension UltraUnification space – a most economical way to generate a Cosmos.

12.1.1 A Qualitative View of Cosmos Spaces

The four levels of Fig. 12.1 are viewed as forming a tetragram that reflects the *structure* of the Physical Reality (being) of the Cosmos. They give a fourfold structure to the Cosmos.

The Cosmos spectrum of spaces are a conceptual abstraction. They are not material. They are therefore *not* created entities. They have no beginning. They have no end. They are timeless. They do not change. They fulfill Parmenides' dictum in his poem *The Way of Truth*:

"There is no change."

They may be viewed as generated from one dimension in the 88 dimension space. From the specification of Cosmos spaces one may create and populate material universes having mass-energy as we have discussed in earlier books.

12.2 Reduction of the UltraUnification Space FRF to One Dimension

We demonstrate this possibility with a simple model example. Assuming the UU FRF has one non-zero dimension (element) a in row 1 of the UU FRF vector with all other components zero we define the vector with d_{c88} components as

$$V = (a, 0, 0, \dots) \tag{12.2}$$

and a *prototype* 88 space-time dimension General Relativistic transformation

$$[T]_{ij} = b_i\ \delta_{i1} + (1 - \delta_{1i})(1 - \delta_{j1})c_{ij} \tag{12.3}$$

where $b_i \neq 0$ and the c_{ij} are numeric. Then T has entries b_i in column 1 of all rows and a matrix c_{ij} in the rows and columns for i, j = 2, 3, … . Consequently a vector

$$V' = TV \tag{12.4}$$

is generated with all components non-zero. The generated vector V' has $d_{cN} = 2^{46}$ non-zero components which become d_{cN} dimensions. This vector can be written as a square array with 2^{46} components[77] where Blaha number[78] N = –35 using

$$N = 9 - r/2 \tag{12.5}$$

The dimension array elements are generated in a vector from one FRF dimension (element). Having the 42 space-time dimension array one can then map to the individual HyperCosmos and Second HyperCosmos dimension arrays.

We can generate the complete set of HyperCosmos and/or Second Kind HyperCosmos dimension arrays from a one dimension element.

---

[77] In agreement with eq. 6.8a.
[78] Negative values of Blaha number N are allowed. They do not appear in the ProtoCosmos model.

Level

1   88 Dimension
UltraUnification
Space

42 Dimension
Full HyperUnification
Space

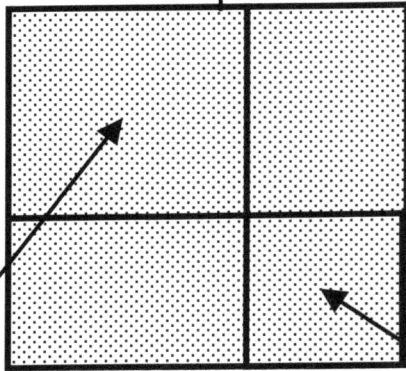

2

10 HyperCosmos HyperUnification Spaces

10 Second Kind
HyperCosmos HyperUnification Spaces

3

10 HyperCosmos Spaces

10 Second Kind HyperCosmos Spaces

4

Figure 12.1. Diagram of the four levels of the Cosmos. They consist of 42 spaces.

13. From Dimensions to Mass-Energy

The development of the set of Cosmos Theory spaces was based initially on a consideration of totally antisymmetric matrices and the Fourier expansion b's and d's of fermion wave functions. The previous chapters developed additional spaces based on unification considerations. They led to the consideration of the generation of spaces from one dimension in an 88 dimension UltraUnification space from General Relativistic transformations in that space.

These considerations were based on dimensions – valueless items with no inherent structure and no mass-energy. They are the primitives of the Cosmos Theory discourse.

13.1 Generation of Dimensions with Mass-Energy from General Relativistic Transformations

In 1978 the author developed a view[79] of the transformation of accelerating coordinate systems and of other coordinate systems related by General Relativistic (GR) transformations. A particle in one coordinate system became a swarm of particles from the viewpoint of other coordinate systems. The essence of this approach was the transformation of fermion wave function creation and annihilation operators under coordinate transformations.

One might take an "extreme" view that all the particles of a system could be generated from a coordinate transformation of *one particle* under a combined General Relativistic and Internal Symmetry transformation. In the case of the 88 dimension UltraUnification space there are only General Relativistic transformations. These transformations filter down to combined General Relativistic – Internal Symmetry transformations for the "smaller" spaces that the 88 dimension space embodies.

In the previous chapters we considered only space dimensions and the effect of General Relativistic transformations on them as they map to dimension arrays of smaller spaces.

In view of the map between dimensions and particles (fermions) that we have seen in previous chapters we take the view that the above considerations apply equally to both.

13.2 Generation of *Particles* with Mass-Energy from General Relativistic Transformations

We now turn to particles. We consider a particle with mass – energy in an 88 dimension UltraUnification universe space-time. By applying 88 dimension General Relativistic transformations that parallel those of the previous section, we can generate

---

[79] S. Blaha, "The Local Definition of Asymptotic Particle States", IL Nuovo Cimento **49A**, 35 (1979).

a system of particles of varying internal symmetries, momentum and energy in sub-universes of the 88 dimension universe.

The result is a new form of unification that can generate the contents of a universe – particles with mass-energy – from one initial particle. The transformations that implement this universe generation process can be alternately viewed as GR transformations in the 88 dimension universe or combined GR – Internal Symmetry transformations in the target universe.

This approach is analogous to the mapping of a particle in an arbitrary coordinate system to a rest frame where only the energy is non-zero.

As a consequence we can view the initial state of a universe (before a Big Bang) as one particle in a certain (extreme) coordinate system. This particle is a set of particles when viewed in other (conventional) coordinate systems using a combined GR – Internal Symmetry transformation.

All in all, Cosmos Theory spaces are unchanging conceptual entities without beginning or end. Universes may be viewed as generated from a single particle, in accord with a Consistency Condition, in the 88 dimension space FRF according to the definition of Cosmos spaces. The simplicity of these concepts recommends them.

The consequences of this view are the regularities in particles and interactions seen earlier in this book.

14. Cosmos Theory and the UST

Cosmos Theory has led us to a new, deeper view[80] of matter, interactions and space-times. The theory develops universes from the Cosmos spectrum of spaces using a Cosmos Consistency condition that fixes the maximum dimension of Physical Cosmos spaces to 18. We can view the development pattern of Cosmos Theory as:

0. Cosmos Spaces
1. Consistency Condition for maximum dimension space
2. Parent Universe(s)
3. Children, Grandchildren, … Universes
4. Our UST Universe
5. Universe Particles
6. Universe Interactions

14.1 Unifying Principle for the Sequences

The seven sequences that appear: Cosmos spaces, Coupling Constants, Gravitation coupling constant, Up-Type fermion masses, Down-Type fermion masses, Vector Boson W and Z masses, and universe masses are all based on Mathematics and Geometry:

1. Cosmos spaces based on Mathematics of totally antisymmetric tensors (geometry of surfaces and volumes) and fermion creation and annihilation operators.

2. Coupling Constants based geometry and spin of fermion creation and annihilation operators.

3. Masses based on geometry and the Cosmos $r = 4$ group of the UST layer.

4. Universe masses based on geometry.

5. Vector boson: W and Z masses based on coupling constants.

We conclude the total Lagrangian for our universe is partly fixed:

1. Cosmos Theory gives spaces and dimension arrays, coupling constants and masses.

---

[80] Blaha (2024i).

2. The UST Lagrangian is almost totally defined by using Cosmos Theory interactions, coupling constants and masses (and scalar and vector bosons); and by Yang-Mills formulations.

The constants in the total elementary particle Lagrangian are thus known except for the possible role of Higgs bosons.

Possible additional particles and interactions remain to be found. The remaining great issue is the determination of scattering matrix elements.

The above sequences of spaces, coupling constants, masses, and universe masses all reflect a fundamental basis in the features of dimensions in Geometry emanating from totally antisymmetric tensors through particles with internal universe shells.

This book shows a deep embedding of the Lagrangian quantum field theory in Cosmos Theory groups and their generator matrices, and in the Quadplex formalism for fermions.

REFERENCES

Akhiezer, N. I., Frink, A. H. (tr), 1962, *The Calculus of Variations* (Blaisdell Publishing, New York, 1962).

Bjorken, J. D., Drell, S. D., 1964, *Relativistic Quantum Mechanics* (McGraw-Hill, New York, 1965).

Bjorken, J. D., Drell, S. D., 1965, *Relativistic Quantum Fields* (McGraw-Hill, New York, 1965).

Blaha, S., 1995, *C++ for Professional Programmers* (International Thomson Publishing, Boston, 1995).

_____, 1998, *Cosmos and Consciousness* (Pingree-Hill Publishing, Auburn, NH, 1998 and 2002).

_____, 2002, *A Finite Unified Quantum Field Theory of the Elementary Particle Standard Model and Quantum Gravity Based on New Quantum Dimensions™ & a New Paradigm in the Calculus of Variations* (Pingree-Hill Publishing, Auburn, NH, 2002).

_____, 2004, *Quantum Big Bang Cosmology: Complex Space-time General Relativity, Quantum Coordinates™ Dodecahedral Universe, Inflation, and New Spin 0, ½, 1 & 2 Tachyons & Imagyons* (Pingree-Hill Publishing, Auburn, NH, 2004).

_____, 2005a, *Quantum Theory of the Third Kind: A New Type of Divergence-free Quantum Field Theory Supporting a Unified Standard Model of Elementary Particles and Quantum Gravity based on a New Method in the Calculus of Variations* (Pingree-Hill Publishing, Auburn, NH, 2005).

_____, 2005b, *The Metatheory of Physics Theories, and the Theory of Everything as a Quantum Computer Language* (Pingree-Hill Publishing, Auburn, NH, 2005).

_____, 2005c, *The Equivalence of Elementary Particle Theories and Computer Languages: Quantum Computers, Turing Machines, Standard Model, Superstring Theory, and a Proof that Gödel's Theorem Implies Nature Must Be Quantum* (Pingree-Hill Publishing, Auburn, NH, 2005).

_____, 2006a, *The Foundation of the Forces of Nature* (Pingree-Hill Publishing, Auburn, NH, 2006).

_____, 2006b, *A Derivation of ElectroWeak Theory based on an Extension of Special Relativity; Black Hole Tachyons; & Tachyons of Any Spin.* (Pingree-Hill Publishing, Auburn, NH, 2006).

_____, 2007a, *Physics Beyond the Light Barrier: The Source of Parity Violation, Tachyons, and A Derivation of Standard Model Features* (Pingree-Hill Publishing, Auburn, NH, 2007).

_____, 2007b, *The Origin of the Standard Model: The Genesis of Four Quark and Lepton Species, Parity Violation, the ElectroWeak Sector, Color SU(3), Three Visible Generations of Fermions, and One Generation of Dark Matter with Dark Energy* (Pingree-Hill Publishing, Auburn, NH, 2007).

_____, 2008a, *A Direct Derivation of the Form of the Standard Model From GL(16) (Pingree-Hill Publishing, Auburn, NH, 2008).*

_____, 2008b, *A Complete Derivation of the Form of the Standard Model With a New Method to Generate Particle Masses Second Edition* (Pingree-Hill Publishing, Auburn, NH, 2008)

_____, 2009, *The Algebra of Thought & Reality: The Mathematical Basis for Plato's Theory of Ideas, and Reality Extended to Include A Priori Observers and Space-Time Second Edition* (Pingree-Hill Publishing, Auburn, NH, 2009).

REFERENCES

_____, 2010a, *Operator Metaphysics: A New Metaphysics Based on a New Operator Logic and a New Quantum Operator Logic that Lead to a Mathematical Basis for Plato's Theory of Ideas and Reality* (Pingree-Hill Publishing, Auburn, NH, 2010).

_____, 2010b, *The Standard Model's Form Derived from Operator Logic, Superluminal Transformations and GL(16)* (Pingree-Hill Publishing, Auburn, NH, 2010).

_____, 2010c, *SuperCivilizations: Civilizations as Superorganisms* (McMann-Fisher Publishing, Auburn, NH, 2010).

_____, 2011a, *21st Century Natural Philosophy Of Ultimate Physical Reality* (McMann-Fisher Publishing, Auburn, NH, 2011).

_____, 2011b, *All the Universe! Faster Than Light Tachyon Quark Starships & Particle Accelerators with the LHC as a Prototype Starship Drive Scientific Edition* (Pingree-Hill Publishing, Auburn, NH, 2011).

_____, 2011c, *From Asynchronous Logic to The Standard Model to Superflight to the Stars* (Blaha Research, Auburn, NH, 2011).

_____, 2012a, *From Asynchronous Logic to The Standard Model to Superflight to the Stars volume 2: Superluminal CP and CPT, U(4) Complex General Relativity and The Standard Model, Complex Vierbein General Relativity, Kinetic Theory, Thermodynamics* (Blaha Research, Auburn, NH, 2012).

_____, 2012b, *Standard Model Symmetries, And Four And Sixteen Dimension Complex Relativity; The Origin Of Higgs Mass Terms* (Blaha Reasearch, Auburn, NH, 2012).

_____, 2013a, *Multi-Stage Space Guns, Micro-Pulse Nuclear Rockets, and Faster-Than-Light Quark-Gluon Ion Drive Starships* (Blaha Research, Auburn, NH, 2013).

_____, 2013b, *The Bridge to Dark Matter; A New Sibling Universe; Dark Energy; Inflatons; Quantum Big Bang; Superluminal Physics; An Extended Standard Model Based on Geometry* (Blaha Reasearch, Auburn, NH, 2013).

_____, 2014a, *Universes and Megaverses: From a New Standard Model to a Physical Megaverse; The Big Bang; Our Sibling Universe's Wormhole; Origin of the Cosmological Constant, Spatial Asymmetry of the Universe, and its Web of Galaxies; A Baryonic Field between Universes and Particles; Megaverse Extended Wheeler-DeWitt Equation* (Blaha Reasearch, Auburn, NH, 2014).

_____, 2014b, *All the Megaverse! Starships Exploring the Endless Universes of the Cosmos Using the Baryonic Force* (Blaha Research, Auburn, NH, 2014).

_____, 2014c, *All the Megaverse! II Between Megaverse Universes: Quantum Entanglement Explained by the Megaverse Coherent Baryonic Radiation Devices – PHASERs Neutron Star Megaverse Slingshot Dynamics Spiritual and UFO Events, and the Megaverse Microscopic Entry into the Megaverse* (Blaha Research, Auburn, NH, 2014).

_____, 2015a, *PHYSICS IS LOGIC PAINTED ON THE VOID: Origin of Bare Masses and The Standard Model in Logic, U(4) Origin of the Generations, Normal and Dark Baryonic Forces, Dark Matter, Dark Energy, The Big Bang, Complex General Relativity, A Megaverse of Universe Particles* (Blaha Research, Auburn, NH, 2015).

_____, 2015b, *PHYSICS IS LOGIC Part II: The Theory of Everything, The Megaverse Theory of Everything, U(4)⊗U(4) Grand Unified Theory (GUT), Inertial Mass = Gravitational Mass, Unified Extended Standard Model and a New Complex General Relativity with Higgs Particles, Generation Group Higgs Particles* (Blaha Research, Auburn, NH, 2015).

_____, 2015c, *The Origin of Higgs ("God") Particles and the Higgs Mechanism: Physics is Logic III, Beyond Higgs – A Revamped Theory With a Local Arrow of Time, The Theory of Everything Enhanced, Why Inertial Frames are Special, Universes of the Mind* (Blaha Research, Auburn, NH, 2015).

_____, 2015d, *The Origin of the Eight Coupling Constants of The Theory of Everything: U(8) Grand Unified Theory of Everything (GUTE), S^8 Coupling Constant Symmetry, Space-Time Dependent Coupling Constants, Big Bang Vacuum Coupling Constants, Physics is Logic IV* (Blaha Research, Auburn, NH, 2015).

_____, 2016a, *New Types of Dark Matter, Big Bang Equipartition, and A New U(4) Symmetry in the Theory of Everything: Equipartition Principle for Fermions, Matter is 83.33% Dark, Penetrating the Veil of the Big Bang, Explicit QFT Quark Confinement and Charmonium, Physics is Logic V* (Blaha Research, Auburn, NH, 2016).

_____, 2016b, *The Periodic Table of the 192 Quarks and Leptons in The Theory of Everything: The U(4) Layer Group, Physics is Logic VI* (Blaha Research, Auburn, NH, 2016).

_____, 2016c, *New Boson Quantum Field Theory, Dark Matter Dynamics, Dark Matter Fermion Layer Mixing, Genesis of Higgs Particles, New Layer Higgs Masses, Higgs Coupling Constants, Non-Abelian Higgs Gauge Fields, Physics is Logic VII* (Blaha Research, Auburn, NH, 2016).

_____, 2016d, *Unification of the Strong Interactions and Gravitation: Quark Confinement Linked to Modified Short-Distance Gravity; Physics is Logic VIII* (Blaha Research, Auburn, NH, 2016).

_____, 2016e, *MoND: Unification of the Strong Interactions and Gravitation II, Quark Confinement Linked to Large-Scale Gravity, Physics is Logic IX* (Blaha Research, Auburn, NH, 2016).

_____, 2016f, *CQ Mechanics: A Unification of Quantum & Classical Mechanics, Quantum/Semi-Classical Entanglement, Quantum/Classical Path Integrals, Quantum/Classical Chaos* (Blaha Research, Auburn, NH, 2016).

_____, 2016g, *GEMS Unified Gravity, ElectroMagnetic and Strong Interactions: Manifest Quark Confinement, A Solution for the Proton Spin Puzzle, Modified Gravity on the Galactic Scale* (Pingree Hill Publishing, Auburn, NH, 2016).

_____, 2016h, *Unification of the Seven Boson Interactions based on the Riemann-Christoffel Curvature Tensor* (Pingree Hill Publishing, Auburn, NH, 2016).

_____, 2017a, *Unification of the Eleven Boson Interactions based on 'Rotations of Interactions'* (Pingree Hill Publishing, Auburn, NH, 2017).

_____, 2017b, *The Origin of Fermions and Bosons, and Their Unification* (Pingree Hill Publishing, Auburn, NH, 2017).

_____, 2017c, *Megaverse: The Universe of Universes* (Pingree Hill Publishing, Auburn, NH, 2017).

_____, 2017d, *SuperSymmetry and the Unified SuperStandard Model* (Pingree Hill Publishing, Auburn, NH, 2017).

_____, 2017e, *From Qubits to the Unified SuperStandard Model with Embedded SuperStrings: A Derivation* (Pingree Hill Publishing, Auburn, NH, 2017).

_____, 2017f, *The Unified SuperStandard Model in Our Universe and the Megaverse: Quarks, ... ,* (Pingree Hill Publishing, Auburn, NH, 2017).

_____, 2018a, *The Unified SuperStandard Model and the Megaverse SECOND EDITION A Deeper Theory based on a New Particle Functional Space that Explicates Quantum Entanglement Spookiness (Volume 1)* (Pingree Hill Publishing, Auburn, NH, 2018).

REFERENCES

_____, 2018b, *Cosmos Creation: The Unified SuperStandard Model, Volume 2, SECOND EDITION* (Pingree Hill Publishing, Auburn, NH, 2018).

_____, 2018c, *God Theory (*Pingree Hill Publishing, Auburn, NH, 2018).

_____, 2018d, *Immortal Eye: God Theory: Second Edition* (Pingree Hill Publishing, Auburn, NH, 2018).

_____, 2018e, *Unification of God Theory and Unified SuperStandard Model THIRD EDITION* (Pingree Hill Publishing, Auburn, NH, 2018).

_____, 2019a, *Calculation of: QED α = 1/137, and Other Coupling Constants of the Unified SuperStandard Theory* (Pingree Hill Publishing, Auburn, NH, 2019).

_____, 2019b, *Coupling Constants of the Unified SuperStandard Theory SECOND EDITION* (Pingree Hill Publishing, Auburn, NH, 2019).

_____, 2019c, *New Hybrid Quantum Big_Bang–Megaverse_Driven Universe with a Finite Big Bang and an Increasing Hubble Constant* (Pingree Hill Publishing, Auburn, NH, 2019).

_____, 2019d, *The Universe, The Electron and The Vacuum* (Pingree Hill Publishing, Auburn, NH, 2019).

_____, 2019e, *Quantum Big Bang – Quantum Vacuum Universes (Particles)* (Pingree Hill Publishing, Auburn, NH, 2019).

_____, 2019f, *The Exact QED Calculation of the Fine Structure Constant Implies ALL 4D Universes have the Same Physics/Life Prospects* (Pingree Hill Publishing, Auburn, NH, 2019).

_____, 2019g, *Unified SuperStandard Theory and the SuperUniverse Model: The Foundation of Science* (Pingree Hill Publishing, Auburn, NH, 2019).

_____, 2020a, *Quaternion Unified SuperStandard Theory (The QUeST) and Megaverse Octonion SuperStandard Theory (MOST)* (Pingree Hill Publishing, Auburn, NH, 2020).

_____, 2020b, *United Universes Quaternion Universe - Octonion Megaverse* (Pingree Hill Publishing, Auburn, NH, 2020).

_____, 2020c, *Unified SuperStandard Theories for Quaternion Universes & The Octonion Megaverse* (Pingree Hill Publishing, Auburn, NH, 2020).

_____, 2020d, *The Essence of Eternity: Quaternion & Octonion SuperStandard Theories* (Pingree Hill Publishing, Auburn, NH, 2020).

_____, 2020e, *The Essence of Eternity II* (Pingree Hill Publishing, Auburn, NH, 2020).

_____, 2020f, *A Very Conscious Universe* (Pingree Hill Publishing, Auburn, NH, 2020).

_____, 2020g, *Hypercomplex Universe* (Pingree Hill Publishing, Auburn, NH, 2020).

_____, 2020h, *Beneath the Quaternion Universe* (Pingree Hill Publishing, Auburn, NH, 2020).

_____, 2020i, *Why is the Universe Real? From Quaternion & Octonion to Real Coordinates* (Pingree Hill Publishing, Auburn, NH, 2020).

_____, 2020j, *The Origin of Universes: of Quaternion Unified SuperStandard Theory (QUeST); and of the Octonion Megaverse (UTMoST)* (Pingree Hill Publishing, Auburn, NH, 2020).

_____, 2020k, *The Seven Spaces of Creation: Octonion Cosmology* (Pingree Hill Publishing, Auburn, NH, 2020).

Blaha-129

_____, 2020l, *From Octonion Cosmology to the Unified SuperStandard Theory of Particles* (Pingree Hill Publishing, Auburn, NH, 2020).

_____, 2021a, *Pioneering the Cosmos* (Pingree Hill Publishing, Auburn, NH, 2021).

_____, 2021b, *Pioneering the Cosmos II* (Pingree Hill Publishing, Auburn, NH, 2021).

_____, 2021c, *Beyond Octonion Cosmology* (Pingree Hill Publishing, Auburn, NH, 2021).

_____, 2021d, *Universes are Particles* (Pingree Hill Publishing, Auburn, NH, 2021).

_____, 2021e, *Octonion-like dna-based life, Universe expansion is decay, Emerging New Physics* (Pingree Hill Publishing, Auburn, NH, 2021).

_____, 2021f, *The Science of Creation New Quantum Field Theory of Spaces* (Pingree Hill Publishing, Auburn, NH, 2021).

_____, 2021g, *Quantum Space Theory With Application to Octonion Cosmology & Possibly To Fermionic Condensed Matter* (Pingree Hill Publishing, Auburn, NH, 2021).

_____, 2021h, *21st Century Natural Philosophy of Octonion Cosmology , and Predestination, Fate, and Free Will* (Pingree Hill Publishing, Auburn, NH, 2021).

_____, 2021i, *Beyond Octonion Cosmology II : Origin of the Quantum; A New Generalized Field Theory (GiFT); A Proof of the Spectrum of Universes; Atoms in Higher Universes* (Pingree Hill Publishing, Auburn, NH, 2021).

_____, 2021j, *Integration of General Relativity and Quantum Theory: Octonion Cosmology, GiFT, Creation/Annihilation Spaces CASe, Reduction of Spaces to a Few Fermions and Symmetries in Fundamental Frames* (Pingree Hill Publishing, Auburn, NH, 2021).

_____, 2022a, *New View of Octonion Cosmology Based on the Unification of General Relativity and Quantum Theory* (Pingree Hill Publishing, Auburn, NH, 2022).

_____, 2022b, *The Dust Beneath Hypercomplex Cosmology* (Pingree Hill Publishing, Auburn, NH, 2022).

_____, 2022c, *Passing Through Nature to Eternity: ProtoCosmos, HyperCosmos, Unified SuperStandard Theory* (Pingree Hill Publishing, Auburn, NH, 2022).

_____, 2022d, *HyperCosmos Fractionation and Fundamental Reference Frame Based Unification: Particle Inner Space Basis of Parton and Dual Resonance Models* (Pingree Hill Publishing, Auburn, NH, 2022).

_____, 2022e, *A New UniDimension ProtoCosmos and SuperString F-Theory Relation to the HyperCosmos* (Pingree Hill Publishing, Auburn, NH, 2022).

_____, 2022f, *The Cosmic Panorama: ProtoCosmos, HyperCosmos, Unified SuperStandard Theory (UST) Derivation* (Pingree Hill Publishing, Auburn, NH, 2022).

_____, 2022g, *Ultimate Origin: ProtoCosmos and HyperCosmos* (Pingree Hill Publishing, Auburn, NH, 2022).

_____, 2023a, *UltraUnification and the Generation of the Cosmos* (Pingree Hill Publishing, Auburn, NH, 2023).

_____, 2023b, *God and and Cosmos Theory* (Pingree Hill Publishing, Auburn, NH, 2023).

_____, 2023c, *A New Completely Geometric SU(8) Cosmos Theory; New PseudoFermion Fields; Fibonacci-like Dimension Arrays; Ramsey Number Approximation* (Pingree Hill Publishing, Auburn, NH, 2023).

_____, 2023d, *Newton's Apple is Now the Fermion* (Pingree Hill Publishing, Auburn, NH, 2023).

REFERENCES

_____, 2023e,*Cosmos Theory: The Sub-Particle Gambol Model* (Pingree Hill Publishing, Auburn, NH, 2023).
qqaa
_____, 2024a, *Cosmos-Universe-Particle-Gambol Theory* (Pingree Hill Publishing, Auburn, NH, 2024).

_____, 2024b, *Fractal Cosmos Theory* (Pingree Hill Publishing, Auburn, NH, 2024).

_____, 2024c, *Fractal Cosmic Curve: Tensor-Based CosmosTheory* (Pingree Hill Publishing, Auburn, NH, 2024).

_____, 2024d, *The Eternal Form of Cosmos Theory* (Pingree Hill Publishing, Auburn, NH, 2024).

_____, 2024e, *The Eternal Form of Cosmos Theory Third Edition* (Pingree Hill Publishing, Auburn, NH, 2024).

_____, 2024f, *Fundamental Constants of Cosmos Theory and The Standard Model* (Pingree Hill Publishing, Auburn, NH, 2024).

_____, 2024g, *Quark, Lepton, W and Z Masses of Cosmos Theory and The Standard Model* (Pingree Hill Publishing, Auburn, NH, 2024).

_____, 2024h, *Geometric Cosmos Geometric Universe* (Pingree Hill Publishing, Auburn, NH, 2024).

_____, 2024i, *Particles and Universes of Cosmos Theory* (Pingree Hill Publishing, Auburn, NH, 2024).

_____, 2024j, *Unification of the Subluminal and the Superluminal in Cosmos Theory* (Pingree Hill Publishing, Auburn, NH, 2024).

_____, 2024k, *The Dawn of Dynamic Cosmos Dimension Arrays* (Pingree Hill Publishing, Auburn, NH, 2024).

Eddington, A. S., 1952, *The Mathematical Theory of Relativity* (Cambridge University Press, Cambridge, U.K., 1952).

Fant, Karl M., 2005, *Logically Determined Design: Clockless System Design With NULL Convention Logic* (John Wiley and Sons, Hoboken, NJ, 2005).

Feinberg, G. and Shapiro, R., 1980, *Life Beyond Earth: The Intelligent Earthlings Guide to Life in the Universe* (William Morrow and Company, New York, 1980).

Gelfand, I. M., Fomin, S. V., Silverman, R. A. (tr), 2000, *Calculus of Variations* (Dover Publications, Mineola, NY, 2000).

Giaquinta, M., Modica, G., Souchek, J., 1998, *Cartesian Coordinates in the Calculus of Variations* Volumes I and II (Springer-Verlag, New York, 1998).

Giaquinta, M., Hildebrandt, S., 1996, *Calculus of Variations* Volumes I and II (Springer-Verlag, New York, 1996).

Gradshteyn, I. S. and Ryzhik, I. M., 1965, *Table of Integrals, Series, and Products* (Academic Press, New York, 1965).

Heitler, W., 1954, *The Quantum Theory of Radiation* (Claendon Press, Oxford, UK, 1954).

Huang, Kerson, 1992, *Quarks, Leptons & Gauge Fields 2nd Edition* (World Scientific Publishing Company, Singapore, 1992).

Jost, J., Li-Jost, X., 1998, *Calculus of Variations* (Cambridge University Press, New York, 1998).

Kaku, Michio, 1993, *Quantum Field Theory*, (Oxford University Press, New York, 1993).

Kirk, G. S. and Raven, J. E., 1962, *The Presocratic Philosophers* (Cambridge University Press, New York, 1962).

Landau, L. D. and Lifshitz, E. M., 1987, *Fluid Mechanics 2nd Edition*, (Pergamon Press, Elmsford, NY, 1987).

Rescher, N., 1967, *The Philosophy of Leibniz* (Prentice-Hall, Englewood Cliffs, NJ, 1967).

Riesz, Frigyes and Sz.-Nagy, Béla, 1990, *Functional Analysis* (Dover Publications, New York, 1990).

Sakurai, J. J., 1964, *Invariance Principles and Elementary Particles* (Princeton University Press, Princeton, NJ, 1964).

Weinberg, S., 1972, *Gravitation and Cosmology* (John Wiley and Sons, New York, 1972).

Weinberg, S., 1995, *The Quantum Theory of Fields Volume I* (Cambridge University Press, New York, 1995).

REFERENCES

INDEX

40, 93, 94, 106, 117

42, 15, 94, 105, 106, ·107, 109, 110, 112, 117, 118, 119, 120, 138

80, 117

88, 15, 117, 118, 119, 120

88 dimension space, 1, 119, 121, 122

anti-symmetric tensors, 19

Asynchronous Logic, 126, 138

baryonic force, 138

Black Hole, 125

bradyon, 39, 42

Casimir force, 21, 22, 139

Casimir vacuum, 21

Cayley-Dickson number, 113

charge, 63, 64, 65

child universes, 21, 22, 23

Classical, 127

Complex General Relativity, 126

confinement, 139

Connection Group Representations, 56, 87

Connection Groups, 1, 53, 55, 56, 70, 71, 72, 74, 87

consistency condition, 9, 21, 22, 23

Cosmic Curve, 24, 25

Cosmological Constant, 126

Cosmos, 128, 137

Cosmos Theory, 4, 130

coupling constants, 27

Creation, 128

Dark ElectroWeak interactions, 10

Dark Energy, 125

Dark Matter, 125, 126

Dark matter/energy, 10

Darkness, 10, 55, 56, 57, 87

Darkness number, 56

deep inelastic, 137

degrees of freedom, 28, 35

dimension array, 1, 3, 5, 6, 7, 9, 20, 21, 23, 24, 25, 37, 38, 39, 40, 41, 43, 44, 45, 46, 47, 51, 53, 54, 58, 67, 72, 74, 75, 87, 88, 91, 92, 93, 94, 95, 103, 104, 105, 106, 109, 117, 118, 119, 139

Dirac, 117

Dirac-like Field Equation, 38

divergences, 136

$e^2/1024$, 27

$e^2/16$, 27

$e^2/256$, 27

$e^2/4096$, 27

$e^2/64$, 27

Eddington, 117

electron, 137

ElectroWeak, 125

ElectroWeak generator matrices, 1, 5

ElectroWeak Interaction $SU(2) \otimes U(1)$ group, 41

Equipartition Principle, 11

Euclid, 25

$e\pi/4$, 139

Fermion Dirac Equation, 77

fine structure constant, 137

fractal, 20, 24, 25

fractionation, 31

FRF, 1, 91, 93, 95, 97, 98, 99, 103, 104, 106, 108, 112, 117, 118, 119, 122

F-Theory, 104, 129

Full HyperUnification Space, 105, 106, 107, 112, 117, 118

Fundamental Reference Frames, 1, 91, 96, 97, 98, 102, 103, 114

gambol, 139

gambol mass, 22, 31, 32

Gambol Model, 4, 130, 139

General Relativistic transformation, 103, 119

Generalized Field Theory, 129

Generation, 17, 125, 126, 129

Generation group, 59, 63, 101

Generation Group, 1, 63, 64, 101, 126
generation number, 56
Generation representations, 56, 83, 84
generations, 63, 64, 65, 66
gravitation constant G, 27
grid, 24
hadron scattering, 139
harmonic oscillator, 136
Higgs Mechanism, 127, 136
higher space, 93, 94, 103
Hilbert, 24
Hilbert curve, 20, 24, 25
Hypercomplex Cosmology, 13, 14, 94, 113
HyperCosmos, 4, 13, 67, 72, 91, 102, 115, 129, 138
HyperCosmos of the Second Kind, 107, 111, 112, 114
HyperUnification, 103, 106, 107, 108, 109, 110, 138
HyperUnification space, 93
HyperUnification transformation, 99, 104, 109
HyperUnification Transformation., 105
interactions, 136
Interpenetrability, 11
ISIS, 138
kT, 21
Layer Group, 1, 61, 65, 66, 127
Layer group representations, 53, 56, 85
Layer Groups, 64, 65, 66
layer number, 55, 56, 57
LHC, 4, 126
Limos, 9, 14, 24, 138
Lorenz group, 19
Maxiverse, 67, 74
Megaverse, 4, 72, 127, 128, 138
$M_{observable}$, 29
nesting, 25
Parent universe, 21
Parity Violation, 125
particle interiors, 34
Pauli matrices, 42

percentage of Dark Matter, 12
Periodic Table of Fundamental Fermions, 1
perturbation theory, 137
Planckian, 139
Plato's Theory of Ideals, 6
pressure of a universe, 21
ProtoCosmos, 4, 129, 138
PseudoFermion, 129
Quadplex, 1, 37, 38, 39, 41, 42, 44, 45, 46, 53, 77, 87, 88, 89, 124
Quadplex Dirac equation, 1
Quantum, 127, 136, 138
quantum computers, 136
Quantum Dimensions, 125
Quantum Entanglement, 126
Quantum Gravity, 125
quark, 136
replicates, 91, 95, 99, 103, 104
scaling, 137
Second Kind HyperCosmos spaces, 1, 88, 112, 113, 117, 118
Second Kind HyperUnification space, 113, 118
SL(2, **C**), 57, 89
SO$^+$(1, 3), 57, 99, 102, 115
Special Relativity, 125
species, 63, 64, 65
species column, 55
spin, 136
Standard Model, 125, 126, 136, 137, 138
Strong Interaction SU(4) group, 40
structure, 5
su(1,1), 13, 113
SU(256), 27, 28, 29
SU(3), 125
SU(4), 39, 44, 49, 50, 69, 81
SU(4) generator matrices, 5
SU(8), 129
substance, 5
superluminal, 104
SuperStandard Model, 4, 12, 127, 128

The Standard Model Generations, 54
time coordinates, 93
totally anti-symmetric tensors, 19
Two-Tier Theory, 37
U(4), 126
U(8), 127
UltraUnification, 129
UltraUnification space, 1, 117, 118, 121
UltraWeak interactions, 11
UniDimension ProtoCosmos, 129
Unification, 4, 127, 128, 129
unification space, 94

Unified SuperStandard Theory, 128, 129, 137, 138
universe, 136, 138
universe shells, 33, 124
UST, 4, 17, 28, 31, 34, 35, 57, 70, 89, 101, 123, 124, 129, 138, 139
UU, 117, 118, 119
Web of Galaxies, 126
α/β, 30
γ-matrices, 19, 25
ν', 30, 32
π, 30

About the Author

Stephen Blaha is a well-known Physicist and Man of Letters with interests in Science, Society and civilization, the Arts, and Technology. He had an Alfred P. Sloan Foundation scholarship in college. He received his Ph.D. in Physics from Rockefeller University. He has served on the faculties of several major universities. He was also a Member of the Technical Staff at Bell Laboratories, a manager at the Boston Globe Newspaper, a Director at Wang Laboratories, and President of Blaha Software Inc. and of Janus Associates Inc. (NH).

Among other achievements he was a co-discoverer of the "r potential" for heavy quark binding developing the first (and still the only demonstrable) non-Aeolian gauge theory with an "r" potential; first suggested the existence of topological structures in superfluid He-3; first proposed Yang-Mills theories would appear in condensed matter phenomena with non-scalar order parameters; first developed a grammar-based formalism for quantum computers and applied it to elementary particle theories; first developed a new form of quantum field theory without divergences (thus solving a major 60 year old problem that enabled a unified theory of the Standard Model and Quantum Gravity without divergences to be developed); first developed a formulation of complex General Relativity based on analytic continuation from real space-time; first developed a generalized non-homogeneous Robertson-Walker metric that enabled a quantum theory of the Big Bang to be developed without singularities at $t = 0$; first generalized Cauchy's theorem and Gauss' theorem to complex, curved multi-dimensional spaces; received Honorable Mention in the Gravity Research Foundation Essay Competition in 1978; first developed a physically acceptable theory of faster-than-light particles; first derived a composition of extremums method in the Calculus of Variations; first quantitatively suggested that inflationary periods in the history of the universe were not needed; first proved Gödel's Theorem implies Nature must be quantum; provided a new alternative to the Higgs Mechanism, and Higgs particles, to generate masses; first showed how to resolve logical paradoxes including Gödel's Undecidability Theorem by developing Operator Logic and Quantum Operator Logic; first developed a quantitative harmonic oscillator-like model of the life cycle, and interactions, of civilizations; first showed how equations describing superorganisms also apply to civilizations. A recent book shows his theory applies successfully to the past 14 years of history and to *new* archaeological data on Andean and Mayan civilizations as well as Early Anatolian and Egyptian civilizations.

He first developed an axiomatic derivation of the form of The Standard Model from geometry – space-time properties – The Unified SuperStandard Model. It unifies all the known forces of Nature. It also has a Dark Matter sector that includes a Dark ElectroWeak sector with Dark doublets and Dark gauge interactions. It uses quantum coordinates to remove infinities that crop up in most interacting quantum field theories and additionally to remove the infinities that appear in the Big Bang and generate inflationary growth of the universe. It shows gravity has a MOND-like form without sacrificing Newton's Laws. It relates the interactions of the MOND-like sector of gravity with the r-potential of Quark Confinement. The axioms of the theory lead to the question of their origin. We suggest in the preceding edition of this book it can be attributed to an entity with God-like properties. We explore these properties in "God Theory" and show they predict that the Cosmos exists forever although individual universes (or incarnations of our universe) "come and go." Several other important results emerge from God Theory such a functionally triune God. The Unified SuperStandard Theory has many other important parts described in the Current Edition of *The Unified SuperStandard Theory* and expanded in subsequent volumes.

Blaha has had a major impact on a succession of elementary particle theories: his Ph.D. thesis (1970), and papers, showed that quantum field theory calculations to all orders in ladder approximations could not give scaling deep inelastic electron-nucleon scattering. He later showed the eigenvalue equation for the fine structure constant α in Johnson-Baker-Willey QED had a zero at $\alpha = 1$ not 1/137 by solving the Schwinger-Dyson equations to all orders in an approximation that agreed with exact results to 4th order in α thus ending interest in this theory. In 1979 at Prof. Ken Johnson's (MIT) suggestion he calculated the proton-neutron mass difference in the MIT bag model and found the result had the wrong sign reducing interest in the bag model. These results all appear in Physical Review papers. In the 2000's he repeatedly pointed out the shortcomings of SuperString theory and showed that The Standard Model's form could be derived from space-time geometry by an extension of Lorentz transformations to faster than light transformations. This deeper space-time basis greatly increases the possibility that it is part of THE fundamental theory. Recently, Blaha showed that the Weak interactions differed significantly from the Strong, electromagnetic and gravitation interactions in important respects while these interactions had similar features, and suggested that ElectroWeak theory, which is essentially a glued union of the Weak interactions and Electromagnetism, possibly modulo unknown Higgs particle features, be replaced by a unified theory of the other interactions combined with a stand-alone Weak interaction theory. Blaha also showed that, if Charmonium calculations are taken seriously, the Strong interaction coupling constant is only a factor of five larger than the electromagnetic coupling constant, and thus Strong interaction perturbation theory would make sense and yield physically meaningful results.

In graduate school (1965-71) he wrote substantial papers in elementary particles and group theory: The Inelastic E- P Structure Functions in a Gluon Model. Phys. Lett. B40:501-502,1972; Deep-Inelastic E-P Structure Functions In A Ladder Model With Spin 1/2 Nucleons, Phys.Rev. D3:510-523,1971; Continuum Contributions To The Pion Radius, Phys. Rev. 178:2167-2169,1969; Character Analysis of U(N) and SU(N), J. Math. Phys. 10, 2156 (1969); and The Calculation of the Irreducible Characters of the Symmetric Group in Terms of the Compound Characters, (Published as Blaha's Lemma in D. E. Knuth's book: *The Art of Computer Programming Vols. 1 – 4*).

In the early 1980's Blaha was also a pioneer in the development of UNIX for financial, scientific and Internet applications: benchmarked UNIX versions showing that block size was critical for UNIX performance, developing financial modeling software, starting database benchmarking comparison studies, developing Internet-like UNIX networking (1982) and developing a hybrid shell programming technique (1982) that was a precursor to the PERL programming language. He was also the manager of the AT&T ten-year future products development database. His work helped lead to commercial UNIX on computers such as Sun Micros, IBM AIX minis, and Apple computers.

In the 1980's he pioneered the development of PC Desktop Publishing on laser printers and was nominated for three "Awards for Technical Excellence" in 1987 by PC Magazine for PC software products that he designed and developed.

Recently he has developed a theory of Megaverses – actual universes of which our universe is one – with quantum particle-like properties based on the Wheeler-DeWitt equation of Quantum

Gravity. He has developed a theory of a baryonic force, which had been conjectured many years ago, and estimated the strength of the force based on discrepancies in measurements of the gravitational constant G. This force, operative in D-dimensional space, can be used to escape from our universe in "uniships" which are the equivalent of the faster-than-light starships proposed in the author's earlier books. Thus travel to other universes, as well as to other stars is possible.

Blaha also considered the complexified Wheeler-DeWitt equation and showed that its limitation to real-valued coordinates and metrics generated a Cosmological Constant in the Einstein equations.

The author has also recently written a series of books on the serious problems of the United States and their solution as well as a book on the decline of Mankind that will follow from current social and genetic trends in Mankind.

In the past twenty years Dr. Blaha has written over 80 books on a wide range of topics. Some recent major works are: *From Asynchronous Logic to The Standard Model to Superflight to the Stars, All the Universe!, SuperCivilizations: Civilizations as Superorganisms, America's Future: an Islamic Surge, ISIS, al Qaeda, World Epidemics, Ukraine, Russia-China Pact, US Leadership Crisis, The Rises and Falls of Man – Destiny – 3000 AD: New Support for a Superorganism MACRO-THEORY of CIVILIZATIONS From CURRENT WORLD TRENDS and NEW Peruvian, Pre-Mayan, Mayan, Anatolian, and Early Egyptian Data, with a Projection to 3000 AD,* and *Mankind in Decline: Genetic Disasters, Human-Animal Hybrids, Overpopulation, Pollution, Global Warming, Food and Water Shortages, Desertification, Poverty, Rising Violence, Genocide, Epidemics, Wars, Leadership Failure.*

He has taught approximately 4,000 students in undergraduate, graduate, and postgraduate corporate education courses primarily in major universities, and large companies and government agencies.

He developed a quantum theory, The Unified SuperStandard Theory (UST), which describes elementary particles in detail without the difficulties of conventional quantum field theory. He found that the internal symmetries of this theory could be exactly derived from an octonion theory called QUeST. He further found that another octonion theory (UTMOST) describes the Megaverse. It can hold QUeST universes such as our own universe. It has an internal symmetry structure which is a superset of the QUeST internal symmetries.

Recently he developed Octonion Cosmology. He replaced it with HyperCosmos theory, which has significantly better features. He developed a fractionalization process for dimensions, particles and symmetry groups. He also described transformation that reduced particles and dimensions to a far more compact form. He also developed a precursor theory ProtoCosmos that leads to the HyperCosmos.

The author showed that space-time and Internal Symmetries can be unified in any of the ten HyperCosmos spaces in their associated HyperUnification spaces. The combined set of HyperUnification spaces enable all HyperCosmos dimensions to be obtained by a General Relativistic transformation from one primordial dimension in the 42 space-time dimension unified HyperUnification space.

At present the author devel;oped the Cosmos Theory that incorporates ProtoCosmos Theory, HyperCosmos Theory, Limos Theory, Second Kind HyperCosmos Theory and HyperUnification Spaces. He has introduced PseudoFermion wave functions and theory, He has related Cosmos Theory to Regge trajectories of spaces, parton theory, Veneziano amplitudes,

Fibonacci numbers and Ramsey numbers. He has calculated an approximation to the difficult R(n,n) Ramsey numbers.

He has developed a Gambol Model that successfully accounts for e-p deep inelastic scattering, fundamental particle resonances, hadron scattering, and the inner structure of particles based on confinement through Casimir forces of ideal gambol gases. The Gambol Planckian Distribution was derived.

He has applied the Gambol Model to particles, universes, and the Cosmos of universes. He showed that the Cosmos may have a distribution of 23 universes corresponding to various Cosmos spaces.

Recently he showed that Cosmos Theory follows from the number of independent asymmetric tensors in a dimension r. He also showed the close parallel between the form of γ-matrices and Cosmos Theory dimension arrays. The closeness suggested that dimension arrays have the same importance as γ-matrices for fermions.

He demonstrated that the pressure of fermions within a space of dimension r balances the Casimir vacuum energy force for 18 dimensions. He showed that $2e\pi = 17.02$ marks the critical point where pressure balances Casimir force, which implies r = 18 is the highest dimension Physical Cosmos space. The dimension $2e\pi$ appears to set the approximate dimension for Cosmos spaces with dimension array size $2^{r+4} \cong (17.02/8)^{r+4} \cong (e\pi/4)^{r+4} \cong 2.13^{r+4}$.

Now he has found the sequence of Coupling Constant values in the Standard Model and UST as well as the sequences of Fundamental Fermion masses. The explication of the form of the fermion Dirac equation based on Cosmos dimension arrays is now apparent.

Recently a complete picture of Cosmos Theory and the UST has developed culminating in the present work.